IN THEIR OWN WORDS

SHORT STORIES OF PENNSYLVANIAN'S IN THE REVOLUTIONARY WAR

By

Edward Leo Semler Jr.

Copyright © 2023 by Edward Leo Semler Jr.

All rights reserved by the author.

First Edition: 2023

Library of Congress Control Number: 2023907081

ISBN: 978-1-7376472-2-5

Printed in the United States of America

City of Publication: Schulenburg, Texas

Cover picture is "Spirit of 76" by Archibald Willard.

Cover Layout by Edward Leo Semler Jr.

To all these veterans and their families

TABLE OF CONTENTS

Introduction	1
Military Service in Pennsylvania	5
Thomas Bull	15
Robert Carr	23
Gabriel Cory	27
Robert Covenhoven	33
David Criswell	37
Cornelius Dailey	43
Franz Dido	51
Peter Dooey	57
Hugh Drennan	61

Leonard Engler	65
Mathias Fisher	71
James Flack	75
William Gill	81
Michael Graham	85
Jacob Grist	91
James Guffey	97
James Guthrie	105
Barnet Hageman	111
Frederick Hain	115
John Hall	121
Joseph Harborn	125
James Hays	129
Jacob Hefflebower	133
George Heinish	139
John Hoge	145
James Huston	151
William Jenkinson	161

James Johnston	167
James Josiah	177
Frederick Kalehoft	181
John Kessler	185
James Knight	191
Thomas Laidley	195
Ephraim Lewis	201
William Lyons	205
Peter Mathews	209
Angus McCoy	213
James Mc Kinzey	223
John Pickens	225
Edward Quigley	229
Conclusion	233
References	235
About The Author	239

Introduction

As I was researching my book "Fighting For Pennsylvania In the Early Years 1763 to 1783" – which is about my Pennsylvanian ancestors who fought in the Revolutionary War - I spent a lot, and I mean a lot, of time reading through war pensions. This was very time consuming. First, there are a lot of them. And second, they are written in cursive with a quill and ink, making them very hard to interpret.

My previous book centered on Captain Thomas Askey and Lieutenant Richard Gunsalus and I tell their stories in detail. And since I only had one of their pension applications I had to rely on other applications, or claims, to hopefully mention them or the unit they were serving in at the time. As I read through these applications, I came across a lot of interesting stories. These random pension claim stories are what makes up this book

Federal pension claims, or applications, came in several stages during and after the war. In 1780 there was an initial pension act established for disabled veterans and widows of fallen soldiers. Later in 1818, 1820, and 1822 it was opened to those who fought more than 9 months in the Continental Army, Navy, or marines. And finally, there was the act in 1832 that opened pensions to everyone else to include the militias and those serving as Indian scouts and spies. But unfortunately, any pension submitted before 1800 was lost in a fire at the War Department.

From my research I found that the claims filed under the 1832 pension act typically offer the best information about a veteran's service. They required a lot more proof of service along with the names of the officers you served under. So applicants were very detailed describing their service, so their claim would be approved.

But unfortunately, it wasn't until some 50 years after the Revolutionary War that a lot of veterans were allowed to apply for a pension. Now in their 70's, and beyond, they had to try and remember what happened almost 50 years earlier when they fought for their country. Sadly, a majority of these claims just mentioned the basics facts of their service, such as their unit, people they served under, and the years. Some are not in first person and the claimant is the veterans' widow or other family member having the right to file a claim. But every now and then I came across a Pennsylvanian that had an interesting first-person account about their service. And some were just unbelievably detailed.

If you think about it, there really had to be a lot of factors that came together for these stories to have survived. First, the veteran had to live into old age to make it to 1818 or 1832 and be able to have their story documented. Second, the claim had to be written legibly by the agent documenting the claim. And third, the veteran had to have the personality to relay more than just the basic details needed to prove the claim. I guess it could also have fallen on the agent to have pulled the information out of the veteran. But I think you'll see in the stories I am about to re-tell that these men had personality, and great memories. And as I relay these stories, I am going to be quoting them just as the veteran presented their story to the claim agent and the agent wrote it down. I'll preface the story with any details that may need explaining so as you read their story, you'll have an idea of what they are talking about. And I'll add "author notes" to help explain anything I feel the reader may be unfamiliar with. But I will not try and correct any historical inaccuracies in their stories. For whom am I to correct someone who was actually there.

So, as you sit back and start to read these tales that happened around 250 years ago, think about what was going on during this time period. There is no United States of America until it is penned in the Declaration of Independence in July 1776, just 13 British colonies surrounded by wilderness. Living is rough and tough. Not only is it a daily struggle to have food to eat but most colonist of the time lived on or near the frontier which posed a threat from Indians pushing back against the creep of European settlements. The added pressure of an authoritative government that overburdened them with taxes and regulations made eking out a livelihood near impossible – causing them to fight for their independence.

Military Service in Pennsylvania

I'd like to give you an idea of what the military framework looked like in Pennsylvania leading up to the Revolutionary War. There were well-established volunteer militias along with a regular contingent of fighting men known as Provincials. These men fought under the British in the French and Indian War, which preceded the Revolutionary War.

When the first engagement of the Revolutionary War happened at Lexington and Concord in April of 1775, it was basically the colony of Massachusetts taking on the British by themselves. When it was apparent that their rebellion was in need of assistance, other colonial militias came to assist. At this point in time the 13 colonies each had militias and were individually governed, to a point, under the overall ruling British umbrella. So, when they decided to band together it became apparent that their militias needed to be unified under a central military leadership system. Enter George Washington as the commanding general of the colonial military and the establishment of the Continental Army in June of 1775.

Pennsylvania units established under the Continental Army were known as units of the Pennsylvania Line. These were basically what the Provincials were back during the Indian Wars, a paid force of enlisted and commissioned men. Enlistments in the Continental Army were typically for a year or more and some were until the end of hostilities.

Now even with the establishment of an army, the colonies still stood up and held control of their own militias. And in Pennsylvania this was a bit confusing, because there were various types of militias. You had a Pennsylvania Militia and you had just plain volunteer groups, known as associators. And these militias and associators were made up of men from individual townships and counties. They provided their own weapons, clothing, food, and usually volunteered for short periods of time – around 2 to 3 months. There was no mandatory military service early in the war and a man was free to get involved in the military or not. These militias and associators were controlled by the county or township they joined. Later, there was also the establishment of Rangers that were controlled by each colony or state. Their service was for one year and they were formed to fight on the frontier.

Most of the duties rendered by members of the Pennsylvania Militia during the war fell into three categories. They were used to augment the operations of the Continental Army, duty on the frontier fighting Indians and Europeans in

Northumberland, Northampton, Bedford and Westmoreland counties, and providing guards for supply depots located in Lancaster, Lebanon, Reading, and at various prisoner of war camps.

To fill the gap needed to move troops rapidly in 1776 a sort of thrown together rapid response fighting group of men known, as Flying Camps, were established. Their purpose was to reinforce New Jersey while the Continental Army focused on the defense of New York. Recruitment was difficult throughout the war, especially in the Continental Army. Not a lot of men wanted to sign up for a year or longer. The Flying Camp enlistment was for only 2-3 months, making them more attractive, and boosted enlistment. Unlike a militia or associators, men of the Flying Camps were paid like the Continental Army, even getting a bonus. As an example, a sergeant would be making 90 shillings a month with a 2 shilling and 6 pence bonus when serving outside of Pennsylvania. Although initially beneficial at the beginning of the war, Flying Camps were disbanded by late 1777.

By this time the war was raging, enlistments were poor, and as a result of these two factors Pennsylvania made military service mandatory. The following is a great detailed description of the mandatory service in the Pennsylvania Militia;

New Pennsylvania laws *"required all white men between the ages of 18 and 53 capable of bearing arms to serve two months of militia duty on a rotating basis. Refusal to turn out for military exercises would result in a fine, the proceeds from which were used to hire substitutes. The act provided exemptions for members of the Continental Congress, Pennsylvania's Supreme Executive Council, Supreme Court judges, masters and teachers of colleges, ministers of the Gospel, and indentured servants. Though, as a practical matter anyone could avoid serving either by filing an appeal to delay their service for a period of time or by paying a fine to hire a substitute. It should be noted, however, that a person serving as a substitute for someone else was not thereby excused from also serving in their own turn. The act called for eight battalion districts to be created in Philadelphia and in each of the eleven extant counties (such as Cumberland County). The geographical boundaries for each district were drawn so as to raise between 440 to 680 men fit for active duty as determined by information contained in the local tax rolls. A County Lieutenant holding the rank of colonel was responsible for implementing the law with the assistance of sub-lieutenants who held the rank of lieutenant colonel. Though they held military titles, these were actually civilian officers not to be confused with the military officers holding the same ranks in the Continental Army. The County Lieutenants ensured that militia units turned out for military exercises, provided the militia units with arms and equipment at the expense of the state, located substitutes for those who declined to serve, and assessed and collected the militia fines. It should be noted that these fines were not necessarily intended to be punitive. Recognizing that personal circumstances might in some cases make it inconvenient or even impossible for a particular individual to serve, the fine system was in part devised to provide money in lieu of service in order to hire substitutes. It also provided an avenue for conscientious objectors to fulfill their legal obligation to the state without compromising their religious convictions.*

The men in each battalion elected their own field officers who carried the rank of colonel, lieutenant colonel and major and these officers were then commissioned by the state and expected to serve for three years. Within each county, the colonels drew lots for their individual rank, which was then assigned to their battalion as First Battalion, Second Battalion, Third Battalion, etc."[1]

Battalions had several companies under them and these companies were broken out into classes each time they were called up to muster. When these new *"classes were called up, each captain would deliver a notice to each man's dwelling or place of business. Under the provisions of the Militia Act, each individual summoned had the right to file an appeal asking that their service be delayed and some successfully avoided service by repeatedly filing appeals. The names of these individuals will be found on the appeal lists. The names of those who actually turned out for muster duty would then appear on company muster rolls listing the men in their new arrangement."*[1]

Contrary to a belief that colonial America was so patriotic that everyone wanted to fight for independence, service in the military wasn't something most men wanted to get involved with. That's why there was a move from volunteer service to mandatory service. There are a host of reasons for this, one of which is the fact that the British had a large following of colonial citizens, known as Tories spread throughout the colonies. These colonists supported and even fought for the British. There was also the problem of a man leaving his family to go off for months at a time, leaving his family to fend for themselves. These were times of subsistence living and fields had to be planted and harvested, animals tended to, homesteads maintained, firewood cut, and a whole host of things to maintain - or your family would starve. Not to mention protecting them from hostile Indians.

Once in military service living conditions were very rough, food scarce, and salaries usually unpaid. You'll read in a lot of the following stories that these men never received money owned to them for their service. This led to a huge recruitment and desertion problem as the war drug on, especially with Pennsylvania Line units. It was common for entire units to desert. Men would be disenfranchised by service life after joining, or would join to get their enlistment bounty payment and then run off with the money.

Pennsylvania tried to nip this in the bud with severe punishment for those caught. Here is an excerpt from a letter dated the 28th of May 1777 of just a few of the examples concerning the punishment handed down for desertion and/or running off with bounty money.

William Day, a soldier in Captn Smith's Comp'y 4th P.R., to receive 150 lashes on his bare back, well laid on, for Desertion, & Defrauding the United States.

Joseph Brooks, a Soldier in Captn Henderso's Com'y, 9th P.R., to receive 300 lashes on his bare back, for Desertion, and twice Defrauding the Public.

Francis Gallige, a Soldier in Captn Tolbert's Comp'y, 2nd P.R., to receive 600 lashes on his bare back, for deserting twice, reinlisting twice, Defrauding the Public, & twice Perjuring himself; - His Pay to be stop'd; to pay what Bounty he may have receiv'd from the Offier who enlisted him last, & afterwards to be return'd to the 2nd P.R.[2]

To get men to join the Pennsylvania Line in early 1780 they began offing officers who joined clothing allowances and all soldiers are given rum, sugar, tea, coffee, tobacco, and soap allowances. Officers that were receiving ½ pay pensions for 7 years after their service were extended to life. This pension was also granted to those disabled in battle. Land grants were also given in appreciation for service. A major general would receive 2000 acres, brigadier general 1500, colonel 1000, lieutenant colonel 750, surgeon 600, major, 600, chaplain 600, captain 500, lieutenant 400, ensign 300, sergeant 250, and a private 200 acres. This land was tax free to the soldier and would be passed on to the heirs of any soldier killed in action.

This was documented in this proclamation by Pennsylvania;

"Being anxious to promote voluntary Inlistment & fill up the Line of Pennsylvania we have concluded to attempt inlisted

Recruits for the War for a Bounty in Land & Specie, viz. 200 Acres of good Land & 3 half Johannes for every able-bodied Recruit, free from the Rupture Lameness or other Disorder, not more than 45 nor less than 18 Years of Age.

No Deserter from the British Army or Navy or Prisoner of War Apprentices or intended Servants to be admitted to enter.

You will also avoid inlisting Sailors & Foreigners, and more especially Frenchmen on any pretext.

Besides the above you are authorized to promise a Suit of Cloaths yearly & Blanket, a Pint of Rum a week with Tobacco, Soap & sundry other Necesaries with Pay & Rations as the Continental Troops - They are to serve under the immediate Command of Gen. Washington.

As an Encouragement to the Sergeant, he will have a Spanish Dollar or 60 Dollars Continental for every Recruit he inlists who passes muster. And the Officer under whose Direction the Inlistment is made will be entitled to 200 Dollars Continental Money for his care & Trouble.

You will be particularly careful not to Suffer any unfair Practices of catching Persons by putting Money in their Pockets, or such like Acts, but inlist them fairly & openly.

Every Recruit, Care being taken that he is quite Sober."[3]

And by 1781 there were still recruitment and morale issues. The Pennsylvania Line had just plain had enough of years of poor living conditions, lack of food, poor enlistment contracts and no pay. In early January in an unprecedented move, they mutinied.

The nearly 2,500 Pennsylvania soldiers at their winter quarters in Morris Town took up arms, demanded that their concerns be met, and planned to march on Congress in Philadelphia. They even threatened to desert over to the British. The uprising was forceful and even violent against their commanding officers. It finally ended with Pennsylvania agreeing to let men who enlisted in the early years of the war under poor contracts the option to be discharged or re-enlist. Some re-enlisted with newer and better contracts, but nearly half of the 2,500 men did not. As a gesture of good will to those men who did re-enlist the Supreme Executive Council authorized a onetime payment to every non-commissioned soldier who had enlisted before 1780 a payment of £9 (9 pounds) state money above their regular pay.

Because a lot of men decided not to re-enlist there was a huge manpower problem with the Pennsylvania Line, which went through an entire restructuring. Eventually half of the restructured Pennsylvania Line was left in Pennsylvania and the other half was sent to fight in the southern theatre of operations in the southern colonies.

Throughout the war county militias also had difficulties recruiting. It was common for them not to be able to muster enough men to go and fight. And as I had previously mentioned a lot of Pennsylvania counties were mustering militias to fight two wars. Not only were they obligated to send men to fight the British, predominantly to the east, but they were also fighting Indians to the west.

The colonist didn't anticipate the war dragging on for over eight years, which wore heavily on the men in its army. Early failed campaigns in Boston, New York, and Philadelphia didn't help with recruitment and morale. Factor in extremely rough living conditions like those famously endured at Valley Forge at the end of 1777 and it's amazing that these men pressed on in service.

As you read the pension applications of these men, you'll see that most are involved with their county's militia. And you'll see that during one enlistment they might be called to fight the British on the eastern side of Pennsylvania and on their next tour they are on the western side fighting Indians.

So, this should have you well prepared to understand what these Pennsylvanian's were experiencing as you read their claims.

Thomas Bull

The story Thomas Bull is about to describe is the Battle of Fort Washington, which took place early in the war. The Continental Army was reeling from defeats around New York City and General George Washington, the commander of the Continental Army needed a win. Although reeling from losses he was trying to rally his army and establish a foothold. In doing this he contemplating the evacuation of Fort Washington which was strategically located on Manhattan Island. The commanding officer of the fort, Colonel Robert Magaw, felt that he could make a stand and hold the fort with the 3,000 or so soldiers stationed there, most from the 5th Pennsylvania Battalion under his command. He had success defending the fort from previous attacks, with less men, over the past few months. So, he felt confident defending it with even more reinforcements. General Washington discussed his options with his senior officers, such as General Putnam, and it was decided to let Colonel Magaw make a stand. If they could hold onto Fort Washington the Continental Army would be able to regroup and hold strategic ground.

Unknown to General Washington and Colonel Magaw was a huge intelligence breach. One of Colonel Magaw's junior officers had deserted to the British Army, taking with him plans for the fort. These plans detailed among other things, sensitive information as to where the cannons were placed. To make matters worse, sensitive letters from General Washington and his staff had also been intercepted by the British. These letters detailed how General Washington was planning troop movements and the overall low morale and lack of discipline of his men and the militias.

On the 15th of November the British decided to attack. Upon arriving, British General William Howe sent a party under the flag of truce to Fort Washington and ordered it to surrender. The British party informed Colonel Magaw that if he did not surrender, every man defending the fort would be killed. Colonel Magaw still feeling as if he held the upper hand with a strong defense, declined to surrender. So, on the 16th of November General Howe attacked the fort from various directions with over 8,000 men and several ships, known as frigates, sailing on the Hudson River.

Here is how Thomas Bull experienced the battle and subsequent imprisonment.

On this eighth day of September 1832 in Chester County PA. Thomas Bull a resident of East Whiteland Township in the county and state afore said aged eighty eight years who being first duly sworn according to law, doth, on his oath make the following declaration in order to obtain the benefit of the act of Congress, enacted June 7th 1832.

That he entered the service of the United States under the following named officers and served as herein stated; to wit.

That in August 1776 he received a commission as Lieutenant Colonel in the Flying Camp, that the Regiment of which he was commissioned a Commander was raised in the County of Chester and was organized at Downingtown in said county, at which place the declarant first took command: - that William Montgomery was commissioned Colonel of said Regiment, but he never appearing to take command, your declarant became in effect Commander in Chief of said Regiment and continued so to be until taken prisoner as herein after stated – that Joh Bartholomew was Major of said Regiment and James McClure, Jacob Hetherly (who died at New York whilst a prisoner), Samuel Culbertson, Benjamin Wallace and ------Pierce were five of the Captains (the names of the other he does not remember) and Hezekiah Davis was one of the Lieutenants: that immediately on taking command the declarant marched said Regiment towards the scene of the war, first moving from Downingtown to Philadelphia, from thence to Trenton, thence to Princeton and through other places of New Jersey of which he remembers Paulus Hook and Bergen, making frequent halts until he arrived with his regiment at North River opposite Fort Washington; that when he arrived at the place, he found Colonel Frederick Watts and Colonel Michael Swobe with their respective Regiments encamped there, the former from Cumberland and the latter from York County of this state; that General James Ewing of York County PA the took command of these Regiments – that he remembered Major Andrew Kilbraith of Watts Regiment and Captain Schmyser and Captain Drit of Swobes Regiment; that these three Regiments under the *direction of General Ewing built Fort Lee opposite Fort Washington in which they all remained until the night of the fifteenth of November 1776, when a detachment from said Regiment put under the command of the declarant was ordered over the river to assist in defending Fort Washington, upon which an attack was expected to be made soon; that Fort Washington was commanded by Colonel Magaw of Carlisle PA under whom were Colonels Rawlings and Cadwallader; that immediately after the arrival of his detachment at Fort Washington on the night of the 15th, a council of war was held by the chief officers, the determination of which was put on paper and placed in the hands of the declarant, with orders to deliver the same to General Washington, who had his quarters about 4 miles west of Fort Lee; that he crossed the river for that purpose and proceeded to General Washington's quarters but the General having rode out, the declarant delivered the papers to General Green and then returned to Fort Washington, by which time it was nearly day break, when he laid down but was soon awakened by the firing of the enemy, which was early in the morning; that he was put on the back of the only horse at the Fort (except one artillery horse) and moved with his Regiment toward Harlem River and after placing it in the order and at the place commanded, he was directed to return to the Fort for other orders; that as he returned a cannon ball from the enemy the force of which was destroyed by striking the limb of a tree fell at the feet of his horse; that on his return he went to the commanding officer when reinforcements of two companies, one under the command of Captain Dean and the other under the command of Captain Hartman, having first arrived from Fort Lee, he was ordered to march them to Harlem River, to be put under the* *command of Major David Lenox, which he did; that when he arrived with them; the Highlanders had landed and driven Major Lenox back to a place called the White House where he delivered his said reinforcements; that he returned partly through the woods to the Fort and was ordered by Colonel Magaw down the North River to call in Cadwallader, to whom he went and delivered his orders, which were immediately obeyed; that he proceeded to the rear of Colonel Cadwallader's command, as they retreated to the Fort and saw Lieutenant Colonel Stewart, who brought up the rear of the forces and was the last man in leaving the field; that the declarant then returned to the Fort and on his way back took up behind him a wounded soldier, who, as he had gone out to deliver his orders to Cadwallader, had passed by the wounded soldier and upon returning carried him back to the Fort; when returning to the Fort he and the wounded man having dismounted; the declarant went into the Fort, finding some of his detachment there immediately went out again and found it safer behind a breast-work near the Fort facing toward Harlem River, being very thirsty, having had no drink since they had left the Fort in the morning (it now being about the middle of the afternoon) he hath water and whiskey brought out to them; that he returned into the Fort after the men had rank when a little English Colonel came to the declarant, he having a double barreled rifle in his hand at the time, "Sir said he, if you will let me have that rifle I will give you 5 or 6 guinea's" the declarant answered that if he would wait till he should lay it down, he might then pick it up; that most of the American forces having been driven into the Fort (it being too small to contain all of them) Colonel Magaw found it necessary to surrender, immediately after which the Americans laid down their arms and the English Colonel took cause to take up the rifle; that the declarant was with his detachment very little during the day having been placed on horseback by Colonel Magaw for the purpose of carrying orders, that by his absence the command of his detachment had fallen upon Major Andrew Kilbraith; that after laying down arms the enemy marched the detachment and his officers to a farm house toward New York, about a mile from Fort Washington, where they were kept from Saturday evening until Tuesday without anything to eat and on Tuesday they were marched to New York where they were put into a meeting house and on the next morning after arriving in New York were ordered out of the meeting house to sign a parole, which they did;*

Authors note – During this era of fighting officers were considered gentlemen and men of honor and their word. They therefore were offered parole which meant they realized that they were prisoners and would not try to escape. Their captors would in turn agree to let them stay within a designated area, in this case the city of New York, to live as if free men.

that the enemy then gave your declarant and his officers part of a house to reside in and furnished them with food, salty beef biscuits, and plenty of coal; that some gentleman, your declarant knows not who, sent him a half-joe

Author's note – a half-joe was a gold coin

when he took boarding for himself and left his officers; that he visited the American soldiers imprisoned in Bridwell almost every day and found many of them dying and dead with cold and want of food but at length orders were given by General Howe for the privates to be discharged on parole (these orders *were understood to be on account of the great number of deaths, General Howe being willing to preserve as many for exchange as possible) that your declarant succeeded in getting his men discharged before any others, that all the officers (prisoners in new York) were then ordered to Long Island to board with farmers, and on the way to the wharf, in obedience to that order, the declarant met Colonel Lowering, Commissary of Prisoners, a Royalist from Boston, who delivered into his hands the billets for the township of Flatlands which he delivered after arriving in Long Island to the officers entitled to receive them, generally two were billeted in a house but one if the in-mates were poor, that he and Colonel Swobe were billeted at a Major Jeremiah Bandabelt, that he remained here until the autumn of 1777 when all the officers on Long Island were put on board of two prison ships laying near New York in which they remained twelve days, when they were ordered back to their boarding houses on Long Island, where your declarant remained until about the 6th of May 1778 when the American Commissary of Prisoners, he thinks his name was Elias Bondinot sent a list of prisoners to be exchanged, the declarant was second on that list and Ethan Allen first, that your declarant soon left New York for his home where he arrived about the 12th of May 1778 having been about one year and nine months in service."*

Author's note – At the Battle of Fort Washington the British suffered over 80 killed and 374 wounded. The Colonist suffered 59 killed, 96 wounded, and almost 3,000 captured. Out of these 3,000 men only 800 would survive captivity at New York and Long Island.

Thomas Bulls pension application was approved and he received $525 a year for serving 21 months as a Lieutenant Colonel.

He passed away the 13th of July in either 1837 or 1839. Both dates are given on documents in his folder.

Robert Carr

Robert's story ties in very nicely with Thomas Bull's in that it also took place at the Battle of Fort Washington. His story is not only interesting because he describes his ordeal; being a prisoner and escaping after the Battle of Fort Washington, but also the fact that he emigrating a year prior to that battle from Ireland and is ready to fight for American independence. And this tour wasn't a one-off. He volunteered for this tour, was drafted once before, and once after this ordeal.

On this 27th day of April 1833 personally appeared in open court before the court of common pleas now sitting Robert Carr a resident of the city of Philadelphia state of Pennsylvania aged seventy eight years on the 25th day of December last. Who being first duly sworn according to law: Doth on his oath make the following declaration in order to obtain the benefits of the Act of Congress passed June 7th 1832.

I was born in Ireland. Came to America in the year 1775 in the ship Alexander sailed from Derry arrived in Philadelphia shortly after my arrival went into Bucks County and hired with *farmer James Mathews, with whom I worked by the month. In the year 1776 in the fall I was drafted in the militia in the company of Captain McConkey and Lieutenant Drake. In a short time marched to the River Delaware above Trenton at I think McConkey Ferry. Marched through Princeton, Brunswick, to Perth Amboy. At Princeton we quartered a few days in the college. At Amboy we did duty for two months where I was discharged. I continued here a few days; when I joined the Flying Camp in the company commanded by Captain Thomas Craig and Lieutenant Craig a relative to the captain. We staid here but a few weeks when orders came and we marched to the North River and up to Fort Lee. Here we were stationed a short time; after which we crossed the North River and proceeded to Fort Washington; here I was engaged until I was made prisoner, when the Fort surrendered, was then marched a prisoner to New York. In New York the officers with whom I was best acquainted, were quartered in a private dwelling. With them I had permission to act as a cook and waiter. After continuing in this situation some time (cannot remember exactly how long) the waiters was about to be reduced to half the number and my fellow cook, being a relation to one of the officers expected he would be retained and I should go into confinement with the other prisoners, concluded I could make my escape. The Captain Craig gave me some hard money, I mingled with the market people, who were crossing into Jersey, paid my passage as one of them, was not suspected and was safely concealed as a market-men, on the Jersey shore at Paulus Hook I then proceeded by secret routes got clear off and got home to Bucks County. This tour was eight months from the time of entry in the company of Captain Craig. The 3rd tour I was drafted into a company* *under command of General Potter, my captain's name I cannot recollect. I marched with several other companies to cross Skipjack Creek and I think Perkiomin, our march was rapid, about 25 miles and arrived near Germantown, we were encamped on the right of the main army near Schuylkill, there to remain until further orders when orders came for us to retreat, so that I was not engaged in the Battle of Germantown. I served out this tour of two months and was discharged.*

Author's note – Robert Carr's pension application was approved and he received $40 per year for serving 1 year as a Private.

There is no date of death in his folder.

Gabriel Cory

In history, a lot of the fighting on the western war front, or frontier, of Pennsylvania seems to get over looked by the more famous battles fought against British regular troops to the east. Gabriel Cory provides excellent insight into fighting against the Indians and their British loyalist allies. His continual drafting by the local militia also shows how often men were asked to fight. Although he may have been getting paid to replace or stand in for someone else in-between his required service, which was a common money-making opportunity. Although he spends several years in the New York militia, he later moves to Pennsylvania and is involved in the savage but little-known Battle of Wyoming, Pennsylvania.

Wyoming, Pennsylvania is in the vicinity of current day Scranton. The battle there in 1778 is also referred to as the Wyoming Massacre due to the huge losses by the Colonists. It's estimated that over 340 colonists were killed there, compared to only 3 Indians and British Loyalists.

You'll have to remember at this time in American history, basically everything west of Philadelphia and New York was contested frontier territory. The British and Colonists had fought the Indians and their allies, the French, there up until the Revolutionary War. And now were fighting the Indians and British Loyalists, known as Tories, there.

On this twenty third day of January AD eighteen hundred and thirty three in open court before the honorable David Scott president and John Coolbough and David W Dingman esquire operate judges of the Court Of Common Pleas in and for the said County of Pike now sitting appears Gabriel Cory a resident of Milford Township in the said county now in the seventy eighth year of his age who being duly sworn according to law doth on his oath make the following declaration in order to obtain the benefit of the Act of Congress passed June seventh AD eighteen hundred and thirty two. That he was born in the town of Goshen, County of Orange and state of New York in the year of our Lord seventeen hundred and fifty four and resided there until the year seventeen hundred and seventy seven, when he moved to the state of Pennsylvania and settled at Kingston on the Susquehanna River in Luzern County where he resided a short time having been driven away by the Indians, and soon after settled in the said county of Pike where he has resided for the last thirty years. That in the year 1776 he entered the service of the United States as a drafted militia man in the company of New York militia commanded by Lieutenant William Stewart, marched to Fort Clinton on the Hudson River in the state of New York and served as a private in the United States service at said fort one month and after performing his tour of duty he returned to Goshen, that late in the fall of the year 1776 he again was drafted for the term of one month and served as private in the New York militia in the company commanded by the said Lieutenant William Stewart and served for the term of one month. During this tour he marched to Fort Montgomery and helped in repairing the said fort which was situated on the Hudson River in the state of New York and after performing his tour of duty returned to Goshen. That in the winter of the year 1776 he was again drafted for the term of one month and served one month as a private in the New York militia in the company commanded by the said William Stewart. During this tour he marched to Toppon Bay on the Hudson in the state of New York and served there one month, after performing his tour of duty he returned to Goshen he further saith that in the spring of the year 1777 he was again drafted for the term of one month and served one month as a private in the New York militia, during this tour he marched to Hackensack on the Hudson River and remained there until his tour of duty had expired. At this time the British Army had possession of the city of New York.

And he further saith that in the fall of the year 1777 he moved to Kingston in Luzern County on the Susquehanna River in the state of Pennsylvania and remained there during the winter of that year, that in the spring of the year 1778 the inhabitants in the Valley of Wyoming received information that the Indians were preparing to make an attack on the people at that place, he was drafted for the term of four months and served in the United States service four months in the Pennsylvania militia as a private. During this tour he was sent by his commanding officer Colonel Butler and Denison up the Susquehanna River about sixty miles for the purpose of removing several families who were in danger of being murdered by the Indians, he together with about thirty other soldiers went up the river and removed the families down the river to Kingston, he was there employed in building forts at Kingston and occasionally was sent out on contouring parties. About the first of July in the last mentioned year the Indians was discovered to be in the neighborhood of Kingston. Our commanding officer Colonel Butler and Denison gave orders to march against the Indians and Tories. After a march of about three miles the Indians met us and gave battle. The American Army counted of four Hundred and forty men, that of the Indians and Tories of one thousand, after a severe and bloody engagement the Americans gave way having been nearly all killed.

Deponent had a brother in this engagement who was taken prisoner and after having the Indians spear slash through him several times he was burned to death. Deponent further saith after the retreat of the Americans in the above mentioned action he retreated on an island in the Susquehanna River where he was kept by the Indians for three days, they having surrounded the island, and while on the island one of his neighbors who had fled there with him was shot down by deponents side, and scalped in his presence when asking for quarter, during this scene deponent lay concealed in a bunch of grape vines. After wondering through the woods five days without any other sustenance but the buds of tree and shrubs afforded he returned to the fort and early the next morning the fort was deserted by the Americans and deponent returned to the state of New York.

Author's note – Gabriel Cory's pension application was approved and he received $26.66 per year for serving 8 months as a Private.

He passed away the 3rd of February 1840.

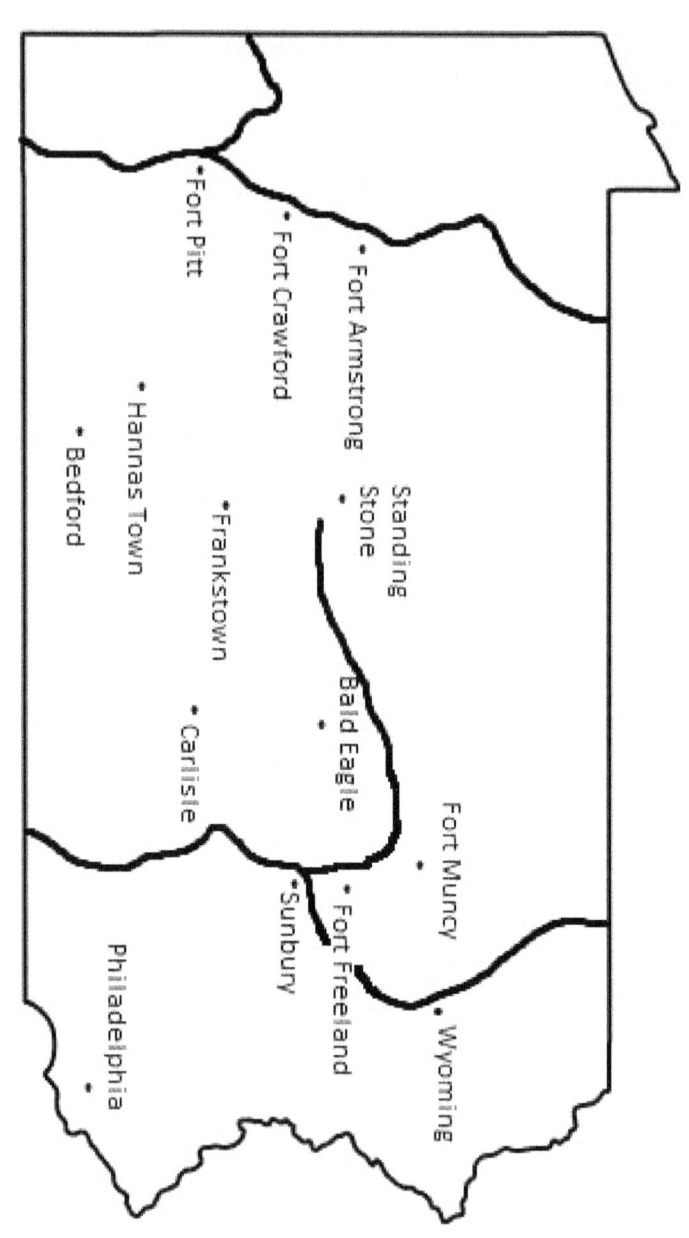

Robert Covenhoven

In reading Robert Covenhoven's pension application he covers an amazing amount of Revolutionary War history. First, he's involved in George Washington's famous crossing of the Delaware River on Christmas Eve and victory over the Hessians. Next, he is on the frontier fighting the Indians and Tories as a scout, spy, and messenger. His exploits of having to travel back and forth through the frontier carrying vital correspondence is an insight into how rough and dangerous three quarters of Pennsylvania was during this time. The only safe haven for those on the frontier where the forts that stretched all the way to Fort Pitt, modern day Pittsburgh. As he delivered correspondence from fort to fort, he was the only line of communication these defenses had. And he had to make his way to and from in a land unmarked by roads or conveniences.

On the fifth day of September Anno Domini one thousand eight hundred and thirty two personally appeared in open court before the honorable Seth Chapman and his associate judge of the Court of Common Pleas now sitting Robert Covehoven a *resident of Mifflin township in the county of Lycoming in the state of Pennsylvania aged seventy six years who having duly sworn according to law doth upon his oath make the following declaration in order to obtain the benefit of the Act of Congress passed June 7th 1832.*

That he entered the service of the United States under the following named officers and served as herin stated. In the autumn of 1776 he volunteered in a rifle company commanded by Captain Cooksey Long in the regiment commanded by Colonel James Murrey of the state of Pennsylvania, he was then living at Fort Muncy in Northumberland County. He marched to Philadelphia and thence to Trenton in the state of New Jersey, where they fought and defeated the Hessians, under the command of General Washington about the twenty fifth of December 1776. He then marched to Princeton where they took three hundred Hessian prisoners. He then marched to Shanks Mills near the mouth of Millstone River, near which place they took from the enemy three hundred wagons laden with plunder and forage. Thence they marched to Morristown where Washington stopped and ordered the volunteers to march to Piscataway then to Short Hills where he the applicant marched under the command of the said Captain Long and Colonel Murray and remained there until they were discharged having been actual serving four months. He then volunteered in the company of Captain William Hepburn in the battalion of Colonel Plunket in the spring of 1778 for the purpose of protecting the inhabitants from the depredations of the hostile Indians. He the applicant being well acquainted with the surrounding country and the location, strength, and mode of warfare of the Indians, he was appointed by Captain Hepburn as leader of the scouting and spying parties, that were constantly on the alert, and watched the movement of the enemy. When the intelligence of the massacre at Wyoming reached the troops stationed at Sunbury Colonel Hunter who had the command at said place inferred orders to captain Hepburn to return to Sunbury, he said applicant carried said express to Antess Fort and they then returned to Sunbury. They there received reinforcements from the eastern counties and returned to Muncy and built a garrison. Colonel Broadhead came to their assistance from Fort Pitt and then the militia were drafted and taken from there. He the said applicant was retained by Colonel Broadhead as a guide to all scouting parties saying that he frequently guided scouting parties to Jenkins Fort and also to the fort at Wyoming and carried express from said places to Colonel Hartly who succeeded Colonel Broadhead at Fort Muncy. After Colonel Hartly had completed the fortifications at Muncy he received orders from Colonel Hunter to march to Tioga Point for the purpose of heading Butler who was carrying off the plunder from Wyoming.

Author's note – There were two Colonel Butler's commanding forces at the Battle of Wyoming. The Colonist were under Colonel Zebulon Butler and the British and Indians under John Butler. Robert Covenhoven is referring to the latter here.

He the said applicant guided the troops under the command of Colonel Hartly to said point where they met the enemy and took from them all their cattle and applicant was then sent (at the peril of his life) with an express to the fort at Muncy to advise them of a party of Indians that had pursued Colonel Hartly to Wyoming and were then in the neighborhood. That he was then sent from the fort at Muncy to Sunbury with express requesting an immediate reinforcement to defend the country and that he again returned to the Muncy fort with the information in reply. He then returned to Sunbury and met Colonel Hartly with troops there. That a reinforcement was dispatched and carried by the said applicant to Fort Muncy – The said applicant was out with scouting parties at various times afterwards, at one time his party were every man killed but himself, and that he alone witnessed the massacre. In 1779 Colonel Hartly left Fort Muncy and joined in Colonel Sullivan's campaign. The fort was left in the care of Captain Hepburn and the said applicant remained with Captain Hepburn and defended the fort and protected the inhabitants against the savages with a small body of men. They received intelligence of a superior force coming upon them and were compelled to retreat carrying with them the defenseless inhabitants to the fort at Sunbury. He was thus in actual service in the state troops of Pennsylvania for a term exceeding two years.

Author's note – Robert Covenhoven's pension application was approved and he received $80 per year, no rank given.

In a 1915 letter from a relative his date of death is given as 1846. It also states he was 90 years, 10 months, and 22 days old.

David Criswell

David Criswell's service spans a large amount of time in the latter part of the war and covers the hardships the frontier county militias faced. His description of service in Cumberland County is typical of what I have read in other pension applications describing the life endured by colonist pushing westward.

You'll notice that he mentions a lot of different forts in his travels. Most of these were just thrown together defenses that were usually not manned on a regular basis. At best they may have been a stone house or fortified homestead. There is little record of most of them except in these pension records.

Henry County Kentucky

On this 11 day of November 1833 personally appeared before me David Adams on of the Justices of The Peace in and for Henry County David Criswell aged seventy two years who being duly sworn according to law doth on his oath make the following declaration in order to obtain the benefits of an Act of Congress passed the 7th of June 1832.

This declarant states that he entered the service of the United States as an Indian spie and served here in stated that he entered the service of the United States as a volunteer soldier under Captain Hugh McLellan of Cumberland County Pennsylvania for a tour of two months that he volunteered sometime in the month of September 1780 but on what day of the month he at this time does not recollect. He marched with Captain McLellan's company from said county of Cumberland to Bald Eagle Creek in the back parts of Pennsylvania in pursuit of some Indians that were committing some depredations on the frontier settlements, on arriving at Bald Eagle Creek he states that the Indians had taken and carried off two of the settlers, John Wilson and his son (see author's notes) *besides committing many outrages upon the property of the inhabitants, Captain McLellan the ordered his men in pursuit of them and our march was directed to an Indian encampment called Chinclacompoose upon the head waters of the Susquehanna River during this fatiguing march in pursuit of the savages he states that he suffered wanting provisions passing through one settled country which was then almost entirely abandoned by whites leaving nothing behind for a soldier to forage upon. Upon arriving at Chinclacompoose the Indians had disappeared, he states that after making several attempts to overtake them and being unsuccessful that Captain McLellan ordered his men back to Cumberland County, and on the way back they met Captain Joseph Brown's company going on the same expedition and after some consultation between the officers it was agreed that both companies should return,* he stated that they returned to Cumberland County and that he having fully and faithfully served out his time was discharged by his said captain. Sometime in the month of November 1780 this declarant states that sometime about the first of March 1781 that he again entered the service as an Indian spie under Captain Thomas Alexander as a volunteer for a tour of two months, he states that Captain Alexander's company was marched from Cumberland to the back parts of Pennsylvania with the view of protecting the frontier settlers and driving the Indians more westerly, he states that they arrived after a tedious march in Nitany Valley on the waters of the Susquehanna where they reconnoitered for a short period, how long he does not distinctly remember – And during this stay in Nitany Valley they were employed in reconnoiter actions, he stated that the company was then marched about 20 miles to Rickett's Fort to prevent an attack upon that fort. Thence they marched to Macommack's Fort and after remaining there a short time they were then ordered to Anderson's Fort, thence back to Cumberland County where he was discharged by Captain Thomas Alexander having fully faithfully and courageously served out his two month tour. He further states that sometime in the month of September 1781 that the Indians again recommenced their depredations on the frontier settlers when Captain Alexander solicited him to volunteer and head a scouting party, he states that he volunteered a third time for a two month tour and started upon a third expedition, during this expedition he states that they marched in a western direction across Stand Stone Mountain and Trish Mountain to Ricketts Fort spieing and pursuing the Indians endeavoring to prevent their depredation made on Macommack's Fort during the abandonment of Captain Alexander's company the Indians ahd become troublesome that the fort was evacuated and who had been there forted had returned to the interior of Pennsylvania for security, he states that the length of time for which he volunteered having expired and he being worn down with fatigue crossing mountains and wondering through a savage wilderness returned with his scouting party to Captain Alexander when they were discharged y him having served out his two months. In the month of September 1782 he states that he was again called into the service as a drafted mountain man, he was drafted in the country of Cumberland and commanded by Captain Archard Coulter his time of draft was for two months, they were then marched from Cumberland County Pennsylvania in a western direction to the frontier settlement on the head waters of the Susquehanna, after crossing Stand Stone Mountain he states that they were stationed at Littles Mill on Stand Stone Creek where they were employed in reconnoitering the country and defending the settlers from attacks by the Indians, another reason for stationing the company at Littles was to guard and protect the mill as it was the only mill on the frontier to supply both the inhabitants and soldiers with bread. In the month of November 1782 Captain Coulter and his company returned to Cumberland County Pennsylvania – and having served out his time for which he was drafted he was discharged by Captain Coulter in writing, his written discharge he states that he has lost many years since. He states that early in the spring of 1783 about the middle of March, as well as he now remembers that he again volunteered under his old Captain Thomas Alexander for the purpose of pursuing some Indians who had broken into the settlements and having committed many depredations upon the property of the settlers, they came to the house of David

Eaton and took his whole family consisting of his wife and five children, during the pursuit after the Indians he stated that they came across David Eaton's wife and two of his smallest children inhumanly butchered, the three largest they forced off with such rapidity they were unable to overtake them, after a tiresome trip after these Indians he states he returned home to Cumberland County being unable to overtake them.

Author's note – David Criswell's pension application was approved and he received $26.66 per year for 8 months service as a private.

He passed away in the spring of 1841 according to a court document in his folder.

I did find mention of a James Wilson and his son John who were kidnapped by Indians during this period. John was sold as a servant and James was adopted by an Indian tribe after he ran the "gauntlet." Later, James was freed and John escaped and walked back to Pennsylvania. David Criswell may have gotten the father and sons name mixed up considering he was remembering back 50 years.

Cornelius Dailey

Here is one of those rare finds in a pension application that blends detail, historical significance, and an interesting story. And all from a ninety-one-year-old man! And his age makes his detail for names and events even that much more amazing.

As you read his story, I just wanted to point out the fact that he was living just 12 miles outside of Philadelphia where major fighting was occurring all around him.

And his account of being near James Monroe when he was wounded at the Battle of Long Island, and who would go on to become the country's 5th President, is a rare find in pension documents.

Something else that you have probably already noticed in previous pension claims is the fact that when these militia men come to the end of their two or three month tour of duty they would pack up and go home, even if it was in the middle of a fighting campaign. This was a problem with the militia throughout the war and General Washington complained about it frequently - that he couldn't count on the militia to stay and finish a fight. But as I stated earlier, the men of the militia had families to take care of and fields to plow and harvest. They could care less at what stage the fighting was in. When their time was up, they went home.

State of New York Monroe County. On this first day of July in the year of our Lord one thousand eight hundred and thirty four, personally appeared in open court before Addison Gardiner Vice Chancellor of a Court of Chancery being a court of recon having common law jurisdiction, a clerk and seal, with power to fine and imprison, held for the state of New York at the city of Rochester, Cornelius Dailey, now a resident of the village of Brockport, in said county of Monroe, of the age of ninety one years, who being first duly sworn according to law does on his oath make the following declaration, in order to obtain the benefit of the provisions made by the Act of Congress passed June 7th 1832.

That he entered into the service of the United States under the following named officers and served as herein stated: That in the month of July 1776, and as this deponent surely believes on the first day of said month he was drafted as a private militia man, out of the militia of the state of Pennsylvania and called out and served under General Robert Doe, Colonel Robert Lewis, Captain Marshall Edwards, 1st lieutenant Richard Whitton, 2nd Lieutenant Samuel Swift, and Ensign John Blake – Regiment not remembered. That when deponent was drafted, he then resided in the Township of Northampton in the County of Bucks, in said state of Pennsylvania – Immediately on being drafted deponent and the company to which he belonged were ordered to march to Amboy, on the eastern shore of New

Jersey near the Rarritan River – opposite Staten Island – And arriving at Amboy they were stationed to guard the coast and prevent the landing of the English troops, who were daily expected. While thus stationed, they occupied part of the time, a large dwelling house as a guard house, belonging to on Cortland Skinner, a Tory, who was then in the service of the enemy – the detachment to which said deponent belonged while at Amboy threw up a redoubt about four feet high, and one and a half miles long; commencing near the mouth of Rarritan River and running east around the eastern point of land near Staten Island. Deponent saw the British fleet from the tops of the houses of Amboy when it first came in sight, and watched it until it passed opposite, between Long and Staten Islands; two ships came up and anchored off about one mile east of Amboy and remained there when deponent left the place, as he supposed to prevent the Americans from attacking a detachment of Hessians which had landed on Staten Island;

Author's note – Mr. Dailey is describing the British fleet as they are coming in to attack New York City, which they would control until 1783.

Deponent stood sentry one night with Judge Jacob Simmons then of Philadelphia County in Moreland Township and who afterwards as this deponent was then informed and believing received a lieutenants commission – went to Fort Washington and was taken prisoner and is since deceased. Deponent also recollects Simon Cruise, John Hilt, William Noll, and many others with whom he was acquainted with he served; but whose names, on account of his age he cannot now recollect. Deponent remained and served at Amboy for at least the term of two months, was discharged and left on the last day of

August or first day of September of the year aforesaid. During the time deponent was stationed at Amboy, the battle on Long Island took place – the Americans evacuated New York and Washington was about returning to White Plains. Deponent received no written discharge whatever – was called up in the morning and marched home with the officers. Deponent was again drafted from the militia of the said state of Pennsylvania as a private while residing in the township and state aforesaid on the first or second day of December in the year aforesaid and marched towards Trenton under the command of Joseph Reid then governor of said state as he believes; Brigadier General John Lacy, Colonel McElroy, Captain Jamison (their Christian names he cannot now recollect) was in the battle of Trenton and stood near Lieutenant Monroe (since president of the United States) when he was wounded – deponent was himself wounded by a dead ball, so called, on his left leg, and by a cutlass on his left shoulder, the scars of which are now plainly visible. Deponent remained in service at this time at least one month as a private and was discharged on the third or fourth day of January 1777, in the same manner as he was at Amboy and marched home with his officers without receiving any written certification of discharge – And this deponent further says that sometime between the twenty seventh and thirtieth day of September 1777, a few days after the British to possession of Philadelphia, and while deponent still resided in the township, county, and state before mentioned, he enlisted as a volunteer in a detachment called the Troop of Light Horse, armed and equipped himself, supplied himself with a horse at his own expense, and under Brigadier General John Lacy, there being no other officer appointed in his company were ordered to watch and guard the roads leading from Philadelphia, and arrest all persons who refused or failed to give a satisfactory account of themselves. Deponent and the company to which he belonged remained doing the like service during the winter following the said September1777. And sometime in the month of March 1778 said troop of Light Horse or part of them in which deponent then was, had a skirmish with the British Light Horse near a place called "Crooked Billet" – about twelve miles east from Philadelphia; in which skirmish about a dozen Americans were killed and some wounded, as near as he can now recollect. Deponent continued upon duty guarded said roads to the best of his recollection until the fifth day of June 1778. Deponent further says that the whole time he served as a volunteer in said Troop of Light Horse was at least eight months, and that he perceived on was to receive during that time twelve dollars per month, but for which Dragoon service this deponent received no compensation whatever, at that time or at any subsequent time and without being discharged from service as volunteer aforesaid. Deponent received an appointment as a wagon master with a captains commission for the term of six months, from Jacob Bennett, then Deputy Quartermaster General under Robert Latis Hooper (as near as he can now recollect the names) which said commission dependent lost, in the manner and under the circumstances here in after mentioned. Deponent was ordered to enlist twelve men. Deponent did so and received twenty dollars county money for each man he enlisted – he then went to said Robert Latis Hooper at Eastown, Northampton County Pennsylvania and got forty nine horses and twelve wagons for baggage loaded with provisions, horse shoes (forty eight of said horses for teams and one for himself to ride) drove them to Morristown *and delivered them to the quartermaster at that place whose name deponent cannot now recollect – At Morristown part of the contents of said wagon were taken out and some more put in, and deponent continued on to White Plains with his company; from there ordered to draw stores from Tarrytown to White Plains. Deponent and his company were thus engaged until the last of September 1778 when James Thomson, then wagon master general ordered deponent and his company to remove to Fredricksburgh. About one month afterwards deponent was ordered from thence to Monmouth, where he remained the remainder of his said term of six months; which expired on or about the fifth of December 1778. Deponent was a regular Continental Dragoon Soldier at the time he was appointed wagon master with a captains commission; And as such continental captain, enlisted men and master general served in a corps of continental teamsters. During which last mentioned term of six months, deponent as such made regular monthly returns to the quarter master general, received money from him, gave receipts, and paid his own men.*

And deponent further states, that while riding on horse back, on his way from White Plains to Tarrytown with said company of baggage wagons he lost his said commission, which he received from said Jacob Bennett, teamster as foresaid together with a considerable sum of money and a pocket book in which it was enclosed and has never since heard of or been able to recover; but does now recollect the whole of its contents except the dates, which are as follows; "Sir, you are hereby appointed wagon master provided you can raise men;- you are to receive captains pay and captains rations when you arrive at headquarters, given under my hand Jacob Bennett, Deputy Quarter Master General."

Author's note – It's kind of sad, but I could find no record of Cornelius Dailey being granted a pension. His claim was denied by the Pension Office in July of 1834 because they said it was defective in some important points.

From the Pension Offices letter, I can only surmise that his claim was denied because he doesn't state the commissioned officers he served under as a teamster.

But in a very long and passionately written response by his attorney, Royal P. Crouse – who states that he is doing the application work for free in gratitude for Cornelius' service during the war – it sounds like there are several other problems. First, there are no living witnesses to his service. And Mr. Crouse states that there is no one left living to give an eye witness account of his service. Mr. Crouse goes on to say that everyone who he has interviewed about Cornelius can attest that it is well known in the community and that he in fact served in the Revolutionary War. Secondly, he addresses an issue with why it has taken Cornelius two years since the pension act to go into law to apply. Cornelius states that he didn't need the money at the time and his pride kept him from applying. But then he was swindled out of his property and money and now could use the pension as he is poor. Mr. Crouse describes him *"as tottering on the nearest verge of eternity – supporting his trembling limbs by means of two canes – and bearing enduring scars (which I have seen) received in the same battle to which late President Monroe was wounded – and worse than all being in a state of abject poverty without a relation whom he can look for aid – he throws himself on the justice of his country for assistance. God grant he may receive it."*

Franz Dido

Although his pension claim was extremely hard to read due to the very bad hand writing of the court clerk, it was well worth taking my time to try and translate it.

He enlisted in a Pennsylvania Line regiment which fell under control of the Continental Army, and he had an enlistment of three years. I find his account of being basically lied to by the recruiter to be true of the time and I'll get more into that in the following paragraph. What is also very interesting his that he was at Valley Forge with General Washington and Lafayette, was a prisoner, served at West Point, and was on hand for the famous mutiny of the Pennsylvania Line.

As I have already mentioned in an earlier chapter, the mutiny was the culmination of years of bad or no pay, horrible living conditions, and unfulfilled promises by their leaders and government. As with Franz, many were tricked into serving with entitlements they never received.

The State of Ohio, Seneca County

On this 3 day of October AD 1833 personally appeared in open court before the Court of Common Pleas (being a court of law) now sitting "Franz Dido" a resident of the township of Clinton said county aged more than seventy five years age he think, and he believes he is more eighty than seventy years old, who being first duly sworn according to law doth on his oath make the following declaration in order to obtain the benefit of the Act of Congress passed June 7 1832. That he entered the service of the United States under the following named officers and served as here in stated. In the winter of 1777-8 I think about the middle of January at McCollister's Town, York County Pennsylvania. I enlisted in the 2nd Regiment of Pennsylvania Troops commanded by Colonel Stewart, Lieutenant Colonel or Major Murray. My company commanded by Captain Rob afterwards by Captain Koby, Then Captain Jacob Stoy, the name of the recruiting officer was Robert Peeling I believe. When I enlisted, I was told by the recruiter that I should be a sergeant of horse but I was soon discovering for I served on foot and as a private. My twin brother of the name Jacob, with several others enlisted at the same time and under similar expectations from the recruiter, but when they found they had been guiled they dissented and earnestly requested me to do the same, but I concluded that my service was much made at that time that I would serve my term which was three years – General Wayne was our general officer under Washington and Lafayette was with me some time. Soon after my enlistment I joined the army at Valley Forge where we remained during the winter. I was in the Battle of Monmouth and next day helped to bury the dead, more killed men that day then living partially or wounded. I was with about 30 others taken prisoner at Newark (and I think it was the *years winter of my service) we was taken to New York on the ice.*

Author's note – Franz is describing the British raid on January 25th 1780 across the frozen Hudson River. They used sleighs to cross over to what is now Jersey City and then moved on to Newark. There they burned down the barracks Franz was staying in and ransacked houses.[5]

We were first in the Sugar House. Nights we was generally locked up in church. I think a Quacker Church. We were prisoners about three months, while in the church we undertook to make our escape. We dug a hole under the wall of the church and under the pavement (I remember I worked with a hogs jaw bone) When we had all things ready, waiting for a dark night a Hessian boy who had previously deserted from the British and joined the Americans and been taken prisoner by the British, (I suppose to make peace) discovered to the British our wood-chuck hole – Soon after we were exchanged, when we rejoined the army I think at Frederickburgh. At the time General Wayne took Story Point I was working at West Point, at the large fort on the hill. Some of the cannon which Wayne took at Story Point we mounted at West Point. The largest I think was drawn up the hill by 16 yoke of oxen and about 60 men with log-rollers.

I wintered at White Plains one winter and at Frederickburgh the remainder I believe. While in the service a baggage wagon upset which I at the lower side was endeavoring to hold up. It fell and my hip was severely injured, however I was very soon on duty again – But the injury was very serious, it was always lame, and more than thirty years ago I became, and have since remained a cripple, the joint being completely destroyed, all which my physician told me was caused by the original injury.

A short time before my term expired there a violent revolt among the soldiers, because they could not get discharged when their term expired, Colonel Stewart was driven out of camp at the point of the bayonet. The soldiers marched toward Philadelphia to show their grievance – General Wayne was with us – I saw him repeatedly in much apparent danger. The soldiers pointing and crying shoot the Damn Rascal. I took no part in the revolt -My term had not yet expired.

However when we got to Princeton (as I believe) and the difficulty was settled and the men getting their discharges, that I myself had but about two weeks more to serve, and thinking at what more the two weeks would be of but little value. Captain Whiteman and another officer gave me a certificate with the rest and we went to the Printing Officer and got discharged. I think mine was signed by General Wayne. But since it was so long ago I cannot for certain who signed it. – I believe it was a printed discharge – I remember the hanging of the spies who had been sent to us by the British.

Autor's note – Franz Dido's pension application was approved and he received $80.00 per year as a private.

He passed away on either the 16th or 25th of June 1841. Both dates are given.

When Franz mentions the hanging of the spies sent by the British at the end of his pension claim, he is referring to men sent by the British during the mutiny. British General Henry Clinton sent men to try and persuade the Pennsylvania soldiers to switch sides and fight for the British, who would grant them pardons and pay them their backpay. The Pennsylvania soldiers didn't take the British up on their offer and eventually turned the men sent by the British to negotiate the deal over to the American authorities, who had the men hung.

Peter Dooey

There are not too many people who can claim that they spoke to General George Washington during the war, but Peter Dooey is one of those who can. His story centers around the Battle of Fort Washington, which should be very familiar to you, the reader, by now. I not only chose his story for his unique interaction with General Washington, but also because he lays out a nicely detailed story of his service.

State of Pennsylvania Cumberland County. On this thirteenth day of August in the year of our Lord on thousand eight hundred thirty two, personally appeared before the Judges of the Court of Common Pleas of said county, now holding a court in for said county. Peter Dooey, a resident of North Middle Township in said county and state, aged seventy three years and upwards, whom being first duly sworn according to law, doth on his oath make the following declaration, in order to obtain the benefit of the Act of Congress passed the 7th of June 1832;

That he entered the service of the United States under the following named officers, and served as herein stated, In the spring of 1776 I marched as a militia man from Carlisle, in said county, in a company commanded by Captain Conrad Snyder, Lieutenant Young and John Crawford, and Ensign Edward Crawford. We marched from here to Philadelphia, and from there to Amboy in New Jersey. The term of this tour was two months.; And that time was nearly ended when we got to Amboy. I then entered into the service again, at the close of the first two months, and joined the Flying Camp; under the same company officers. The Flying Camp was commanded by Colonel Frederick Watts, the father of the late David Watts Esquire of Carlisle, and Major Andrew Galbreath, formerly of Cumberland County – This engagement was for six months. From Amboy we marched to Fort Lee, on the North River – I enlisted to work at that fort for some time. General Greene commanded at the fort. A few days before our six months were up, I was drafted and went across the river to Fort Washington, which was under the command of Colonel Magaw – Major Galbreath and Captain Snyder where also sent over – I crossed to Fort Washington on the 15th of November 1776, and on the 16th of November 1776, the fort and garrison were captured by the British Army, and I and Major Galbreath and Captain Snyder taken prisoner. We were riflemen, and were posted on the upper north side of the fort, and were opposed to the Hessians. The British vessels had gone up the North River before I went over to Fort Washington. One of them got past on the Chevoax de Friez and we wanted to go try to take her, but the officers were not agreed to it.

Author's note – Chevaux de Frise is a defensive obstacle which can be placed in the water and which one of the British vessels had apparently gotten hung up on.

Fort Lee was taken two days after Fort Washington, and on the day it was taken, my time of six months would have been out – I was taken on Saturday; on Sunday the prisoners were taken to New York. We were in close confinement, as prisoners for seven weeks before I was discharged on parole, I suffered very much for want of good water and provisions – I was discharged, on parole, and came from there, through Philadelphia, home – it took me a month to get home; being unable to travel from weakness and hunger – I well remember the taking of the Hessians at Princeton while I was a prisoner in New York; and of English soldiers saying they must now treat us better, or their men would be very badly treated too – At the time of the Battle of Princeton, I was on my way home – As General Washington was marching to Morristown, we met him, and I spoke to him about our situation; and he directed us to go on to Philadelphia. I was not exchanged until the war was over. I never got any of my pay except for one month. I got no discharge in consequence of being taken prisoner.

Autor's note – Peter Dooey's pension application was approved and he received $31.33 per year, for 9 months and 12 days of service as a private.

There is no indication of his date of death in his folder.

Peter Dooey states that he was paroled, but not exchanged until the end of the war. As I have previously stated, officers were usually paroled and required to stay within the confines of their captors-controlled area. Enlisted men in some cases were also paroled if ill or wounded and sent home, on the condition they would not join the fight again until exchanged. Another gentlemen's honor agreement of the times.

Hugh Drennan

Hugh Drennan's nicely detailed pension application covers almost the entire war from 1776 to 1783 and mixes some of the early fighting done by the Pennsylvania Line, the militia, and as a wagon master. He participates in a variety of campaigns such, as The Battle of Three Rivers - all the way up near Quebec, The Battle of Crooked Billet – in which he receives three wounds and is captured, and as a wagon master - due to his wounds limiting his service.

Declaration in order to obtain the benefit of the Act of Congress passed June 7th 1832. The Commonwealth of Kentucky, Fleming County, on this 5th day of March in the year 1834 personally appearing before the honorable circuit court for the state and county aforesaid Hugh Drennan a resident of the aforesaid county and state aged 72 years who being duly sworn according to law doth on his oath make the following declaration fourth – That he enlisted in the army of the United States in the month of January 1776 in Cumberland County Pennsylvania in the town of Shippensburg in the company *commanded by Captain Abraham Smith of the Pennsylvania Line Sixth Regiment commanded by Colonel James Irvine for twelve months, he marched to Carlisle, thence to Philadelphia, thence to the city of New York, thence by way of Albany to Lake George, thence to Ticonderoga, thence t Lake Champlain, thence to the St. Lawrence River, thence to the Battle of Three Rivers, which he was in, when the American Army commanded General Thompson or General Sullivan was defeated, thence the American Army retreated up the St. Lawrence River and marched to a place called Crown - Point and built Fort Harttey and from thence to Mount Independence near Ticonderoga and wintered; from thence by way of Fort Ann and Fort Edward situated in a mountainous and wilderness country, to Philadelphia, and remained at least four weeks, and from thence back to Carlisle in Pennsylvania and received an honorable discharge signed by Abraham Smith in the month of March or April 1777, made the tour of duty in at least 13 months while discharge has been since lost or misplaced so that he cannot nor find it, and in the month of April 1778 he again volunteered in the county of Cumberland for six months, the company was commanded by Captain William Findley of the Pennsylvania militia, the regiment commanded by Abraham Smith, his former captain, and marched to a place called the Cooked Billet near Philadelphia and fought the enemy and received three wounds, one on the head from a horseman's sword, one from a musket passing through the fleshy part of his thigh, and on the hand from a sword, and was taken prisoner on the first day of May 1778 by the British army and was carried to Philadelphia and kept six days and then made his escape and returned to his company which was stationed near Doylestown Pennsylvania at a place called Bowman Hill and which place he was kept until he could return home and was permitted to do so by his captain, and owning to a very weak state of health he was unable to reach his residence in Cumberland County until July, making the tour at least three months, he further states he was unable to labour or do the duty of a soldier during the year1778 and further that in the month of May 1779 he joined the army as a waggoner and drove and acted as such until he reached a place in North Carolina called the High Hills of Santee and in the summer of 1780 was appointed by General Steuben forage and wagon master and continued as such at least 8 months when he resigned and returned to Philadelphia, making in all at least 25 months – that he served in the foraging capacities and engaged as a waggoner again until the end of the war and was discharged at Boston in the winter of 1783 by Colonel Comtsammier.*

Author's Note – Hugh Drennan's pension application was initially denied because they said his dates for his initial service into Canada were wrong and that service as a waggoner was not allowed. But it must have been later granted because his widow, Margaret, files a claim in 1848 requesting it be allowed to be passed to her. And she states that he was receiving $53.33 per year from the 1832 pension act. Her pension, which was approved in 1848, states it is for Hugh's service as a private. So, I will assume his claim was eventually approved as a private, and not as a waggoner, and that he received $53.33 per year, for service as a private.

Margaret Drennan states in a letter to recover her deceased husbands pension benefits that he passed away the 26th of December 1842.

The Battle of Three Rivers saw 2000 Colonists go against a much smaller British force of 1000 men. But the colonists limped away with 4 times the casualties and over 230 men captured.

He is wounded at the Battle of Crooked Billet, which was previously mentioned in Cornelius Daley's pension Application. The battle was fought on the 1st of May 1778 and this matches the day he is taken prisoner.

Leonard Engler

Another finely laid out pension claim detailing a wide variety of militia assignment. One of the major taskings of the militia was guarding prisoners, which is nicely detailed in this claim. He also touches on the infighting that went on between the Pennsylvania Line and militia troops that is well documented throughout the war and I cover it in detail in my pervious book on the Revolutionary War. I found Leonard Engler's account of them bumping heads very typical.

Another subject he touches on is that of standing in, or substituting, for another person. I mentioned this previously in this book and it was very common. Men would pay to have someone stand in for them or family members would freely stand in for each other. The minimum age for a substitute was 18 years of age but this was waived if a family member substituted. I found in my research it was common for boys who were under the age of 18 to substitute for their fathers.

And lastly, he gives another nice first-hand experience with George Washington, which I am always exciting to come across.

On this nineteenth day of April in the year of our Lord one thousand and eight hundred and thirty three personally appeared before the Court of Common Pleas of Northampton County in open court Leonard Engler a resident of the Township of Plainfield in the County of Northampton and Commonwealth of Pennsylvania aged seventy six years and upwards who being first duly sworn according to law doth upon his oath make the following declaration in order to obtain the benefit of the Act of Congress passed the seventh day of June A.D. one thousand eight hundred and thirty two – That he entered the service of the United States in the end of September A.D. 1777 being drafted at is then place of residence in the Township of Plainfield in the County of Northampton and state of Pennsylvania, to serve in a company of militia commanded by Captain William Henry Lawall – That as near as he can tell he was then about twenty one years of age – That he proceeded from home immediately to the house of Captain Lawall in Bethlehem Township where the company mustered – They were then marched to Easton where they were joined by a company of militia commanded by Captain Jacob Weygandt – There were at this time lying in the Easton Jail about 160 prisoners composed of British, Hessians, and Tories as deponent was told brought from Long Island and there abouts, of these prisoners the two companies of Captain Lawall and Captain Weygandt took charge and (deponent being still with his company) immediately proceeded to Lancaster with them, where they safely arrived and lodged them in jail. They then returned as far as Allentown, Lehigh County where they remained for eight days, at which place a large body of militia and regulars lay at the time – At this place a fracas occurred by moon light between the regulars and militia, which but for the timely interference of the officers would have led to serious consequences, and in the course of which Captain Foarseman of the militia was shot by one of the regulars along the cheek and nose with Buck Shot and at the moment he fell this deponent was not three rods from him. He afterwards however recovered. When deponent left Allentown with his company, he thinks the whole body of troops went along – They proceeded to White March about 12 or 15 miles from Philadelphia (the city at that time being in the possession of the enemy) At this place, White Marsh, deponent frequently saw General Washington passing through the camp and mounted upon a sorrel horse the largest that deponent ever saw. He frequently saw him ride around directly after sunset attended by his officers and eight men mounted on black horses, four before them and four behind. After lying there for a few weeks it was said that five hundred militia were wanted to go to Red Bank Fort. Colonel Siegfried who commanded deponents regiment insisted that the whole regiment should go, and there was some little misunderstanding and difference among the officers in relation to this, but about 10 o'clock upon the same morning it commenced raining very hard and rained for three days so that the troops could keep no fire to bake or cook, and during the whole of that time deponent had nothing to eat, his flour being soaked with water and his meat raw. For two nights deponent slept under a wagon is a hole which he had dug with the dirt banked up around it to turn off as much as possible the water which was running in torrents. At this time too, he had the additional privation of being without a coat, some of the soldiers having stolen it from him. On the third day of this storm of rain they heard a great explosion which the men at the time allowed was Red Bank Fort – but deponent does not know what it was. Not long after this, the whole body of troops moved down nearest Philadelphia to Farmers Mill, where they encamped on a Stoney Hill which the men called Bunker's Hill. Soon after Washington moved the regulars across the Schuylkill leaving the militia behind. The militia about this time increased very fast. A report reached the camp that the enemy were marching out of Philadelphia and making preparations for a attack. The militia were numerous and in high spirits and every preparation was made with the utmost alacrity to receive the enemy – deponents company staid several days over their time in expectation of sharing in a battle, but finding there was little or no probability of it, they returned home in the month of December after an absence on duty of two months and three or four days. Deponent further states that so far as his knowledge extends there is but one surviving witness to the above described service to wit Henry Lause of Forks Township in this county and whose deposition is here to annexed.

And the deponent further says that he was again drafted in the militia and left home in the month of April A.D. one thousand seven hundred and eighty one in a militia company commanded by Captain Abraham Horn of Easton upon an expedition against the Indians who had been committing depredations upon the inhabitants in the neighborhood of the Wind Gap and Water Gap, and at that time in this country they had attacked a family of the Deckers and several other families massacring all whom they got into their power. The militia *proceeded to the neighborhood and were stationed in scattering bodies of men from the Wind Gap along to the river and reaching above Vanatta's, a distance of forty five or fifty miles. How many men were there deponent cannot say, or who was first in command – The highest officer he saw was Colonel Jacob Stroud. Deponent having been only drafted for two months, at the end of that time he agreed to serve and did serve another tour of two months as a substitute in the place of one John Shaeffer of Nazareth in this county. The company still being under the command of Captain Horn. At the end of his second tour the company being under the command of Captain Jacob Balliet. Deponent remained again as a substitute for some person whose name he has forgotten. The way in which it happened that Captain Horn commanded a company the second tour was that he was a substitute for some other captain and his company was made up of some new drafts and a number of the men who had come out for the first tour and remained as substitutes – The company in which deponent served the third tour under Captain Balliet was composed in the same way. At the end of deponents third tour another draft of militia came up, but the inhabitants, thinking it unnecessary for them to remain, they returned almost immediately. Deponent returned at the same time after an absence in the service of the United States as above detailed of not less than six months. During the first two – month tour deponent with a few others saw five Indians standing on the hill above Deckertown – Deponent, five of the Decker family, and a man from below Nazareth started after them with four dogs and followed them till dusk but were not able to overtake them. Deponent never received a discharge from the service and believes that none were given to any body. As to witnesses deponent says that Jno Reichner of More Township is alive and able to testify to four months of the last mentioned service. All the other persons in deponents neighborhood who accompanied him at that time are dead.*

Autor's note – Leonard Engler's pension application was approved and he received $26.66 per year, for 8 months service as a private.

There is no date of death given, but his attorney requests a speedy decision because Mr. Engler will not live much longer.

Mathias Fisher

In this exceptional claim we start with Mathias Fisher marching all the way across the state of Pennsylvania to fight in New Jersey. Then he march's all the way back across the state again to be discharged. He then details an amazing story about the failed expedition into hostile Indian territory, led in part by Colonel Archibald Lochry, in 1781. An estimated 37 members of Colonel Lochry's party were killed in the initial attack on them and 66 taken prisoner, with over half of the prisoners later dying. The fact that Mathias Fisher lived to get back is amazing in itself, but to have it documented in his own words is a rare find.

To give you a little background on what you are about to read, I'll briefly summarize the expedition he was on. In August of 1781 Mathias Fisher was part of a group of soldiers traveling into present day West Virginia and Ohio. They were part of a bigger party of men planning an attack on the British in Detroit. As Mathias Fisher's party was traveling by water, they got separated from the other group they were traveling with.

What you are about to read is referred to as Lochry's Defeat or Lochry's Massacre.

On this 20th day of November in the year 1832 personally appeared in open court before the honorable John Young now sitting Mathias Fisher a resident of Ligonier Township Westmoreland County in the state of Pennsylvania aged 74 years who being first duly sworn according to law doth depose and say in order to obtain the benefit of the Act of Congress of 7 June 1832; That about the first day of January 1777 as declarant believes he was drafted as a militia man in the company of Captain Shannon and marched in the detachment commanded by Colonel Archibald Lochry from Westmoreland County in Pennsylvania to the state of New Jersey at Morristown thru to Brunswick and Amboy we went to the American lines, and were stationed at Raway near Woodbridge, which place we left I the spring and the whole detachment returned I think in April, being discharged. – During the time we lay at Raway there was one or two small affairs with British foraging parties – In the month of August seventeen hundred and eighty one the detachment marched from Fort Ligonier in Westmoreland County Pennsylvania under the same captain and the same Colonel Lochry to Wheeling in Virginia, and then the whole detachment embarked in boats commanded by Colonel Lochry, we landed below the mouth of the Big Miami at the mouth of a creek and while we were landing the Indians fired upon us and finally took the whole detachment prisoner. Colonel Lochry was killed after we surrendered, there was about a hundred in the detachment as well as the declarant recollects when we were attacked. The declarant was kept a while in the Indian town and about Christmas was taken to Detroit and given up to the British. I was kept at Detroit until spring when I was taken to an island about forty five miles above Montreal - about 20 acres in the island on which then was a fort and a great many American prisoners – about the 13th of July 1782 as I believe myself and four other American prisoners to wit Ezekial Lewis, Samuel Murphy, James Dougherty, and George Baily, the two last of whom men where at Wyoming made our escape in the night by passing the guard and constructing a small craft out of drift wood found on the island on which we drifted about five miles before we got to land and that was on the shore when we concealed ourselves during the day and at night got a canoe and crossed to the American shore and steered for Lake Champaign which we landed at the mouth of Otter Creek and crossed the lake on a raft, which we thought was about for miles broad at that place, we thus went to Pittsford and on to Allentown where Governor Chittenden lived and gave us a pass, to Newbury the headquarters of General Washington from thence to Easton Pennsylvania from where I returned to my home in Westmoreland County Pennsylvania which I reached about the 12th of September.

Autor's note – Mathias Fisher's pension application was approved and he received $50.00 per year, for serving 15 months as a private.

He passed away the 17th of February 1834.

As for the men that escaped with Mathias from the island; Ezekial Lewis was captured with him in August. Samuel Murphy is unknown – but there was a Patrick Murphy captured with him in August and he may have gotten the first name incorrect in his old age. Next, he states that James Dougherty and George Baily were captured at the Battle of Wyoming, in which you have read about in previous pension applications.

As Mathias Fisher and his group crossed Lake Champaign they entered Vermont, where Governor Chittenden gave them a pass which let them basically travel unmolested back home.

James Flack

Following up on the previous great story of being captured by the Indians comes another amazing story of service and subsequent capture by the Indians. James Flack's story is a bit more detailed than the previous one and you will see that he was one tough dude!

On this twenty second day of May in the year 1834 personally appeared in open court before the Court Of Common Pleas of Westmoreland County Pennsylvania James Flack a resident of Ligonier Township in the county and state aforesaid aged eighty one years who being first duly sworn according to Law doth on his oath make the following declaration in order to obtain the benefit of the proposition made by the Act of Congress passed June 7th 1832;

That in the year 1777 this declarant entered the army in Westmoreland County Pennsylvania as a volunteer in a mountain company commanded by Captain John Hickson in the month of September of said year, we were stationed at Palmers Fort in Ligonier Valley, during the period of

Declarants service at Palmers Fort he volunteered with others of a company to go in pursuit of a party of Indians who had committed a number of depredations in the neighborhood, we overtook the Indians in the night at a place called Blanket Hill three miles from where Kittanning now stands, we surprised and killed the whole party except one who was supposed at the time to have been the notorious Simon Girty.

Author's note – Simon Girty was a colonist who had assimilated into Indian culture and acted as an interpreter between the British and the Indians.

Declarant remained in the service at Palmers Fort for three months when he returned home.

In the month of February or March 1778 this declarant entered as a volunteer in a company commanded by Captain Robert Knox, we rendezvoused at Fort Ligonier in Westmoreland County from thence marched to Pittsburg where we joined General Hand who commanded an expedition intended to go out against certain towns in the Ohio country, we marched about sixty or seventy miles into the Indian Country and destroyed two Indian towns. The second town was fifteen miles beyond the first. This declarant was one of sixty volunteers by whom its destruction was effected, while engaged in this service we were commanded by Colonel Pumroy, declarant was on of the mountain volunteers, Colonel Crawford, Colonel Mason and Colonel Stinson were attached to the expedition. The declarant was absent on this tour two months.

In the month of May in the year 1780 having gone to Kentucky he entered as a volunteer under Captain George Riddle who commanded a small fort called Licking Station situated on

Licking River in Kentucky, the fort was about forty miles from Limestone. The garrison consisted of about forty men and boys. On the 24th of June 1780 before sunrise, we were attacked by a large force of British and Indians amounting as the declarant was afterwards informed by one of the enemys artillery men to two hundred and fifty British and eight hundred and fifty Indians. The attack and defense was continued until about two hours before night. The enemy then brought forward a piece of cannon, which cut our picketing and compelled us to surrender, declarant lost a horse and two rifles, one of the rifles had belonged to declarants brother who was shot by the Indians while standing sentry at the fort two days before the action. By our articles of capitulation, we were to be retained by the British as prisoners of war. But no sooner had we surrendered then we were delivered to the Indians with the exception of two individuals, Captain John Hinkson and Enos McDonald. The Indians bound this declarant and marched him to where Cincinnati now stands and from thence to beyond Detroit. The Indians then brought declarant to Detroit and delivered him up to the British. I came to Detroit on the 28th day of July having been with the Indians from the 24th day of June. Declarants health being injured from the severe treatment he had received he was allowed the privilege of the yard of the barracks. After I had recovered my health, I was confined in jail for three weeks for chastising a refugee for insulin; at the end of three weeks I was liberated from jail and had the privilege of the barracks as formerly. A few days after my release I went into the Smith Shop with a fellow prisoner. I there saw two rifles which I had known before they were taken from Ligonier Valley by the Indians. We damaged the sides of the rifles that they might be of no service to our oppressors.

For this act thirteen of us were placed aboard a prison ship and sent down to Niagara. Below Detroit we made an attempt to take the vessel but one of our own party proved a traitor, our plan was detected, and we were all secured and bound. At Niagara I was offered four shillings Sterling a day to assist in making gun carriages, this I refused and was again imprisoned for what they called my obstinance. I remained in jail until about the first of May 1781. I was then taken to Montreal where I remained in prison until the last of June or the first of July when through the influence of Michael Rugh of Westmoreland County then a prisoner at Montreal, and afterwards a representative of said county in the General Assembly of Pennsylvania. I was taken out of jail and let to work for Captain Grant a British officer. I worked on an island opposite Montreal called Grant Island and was allowed one shilling a day York money. After working some time I received a half-joe from Captain Grant, with this I bought two compass dials, some cheese, a bottle of rum, and a pair of moccasins. Having made these preparations this declarant took advantage of a dark night and in the company with five other prisoners crossed the St. Lawrence in a canoe and made the best of our way to Vermont, and after ten days painful traveling we arrived at General Baileys in Vermont. Genera Bailey gave us a pass to Boston and a recommendation to General Hancock, from Governor Hancock we got a pass to Philadelphia and an order for then days provisions. At Sussex in the Jerseys I parted from my companions and came into Bucks County Pennsylvania where I arrived in the latter part of the month of August 1781. I remained in Bucks County for some time among my relatives to rest after my fatigue and arrived at my fathers in Cumberland County in the month of October 1781.

Autor's note – James Flack's pension application was approved and he received $56.66 per year, as a private.

There is no mention of his date of death in his folder.

William Gill

Although his pension claim is relatively short compared to others that I have included in this book, it is still very interesting. Primarily because of the amount of times he was shot, wounded, and just kept on fighting.

State of Pennsylvania, County of Mercer, On this sixth day of September A.D. 1832 personally appeared before Alexander Brown, a Judge of the Court of Common Pleas in and for the county aforesaid William Gill aged 82 years, a resident of Wolf Creek Township, Mercer County Pennsylvania, who being first duly sworn according to Law, doth, on his oath make the following declaration in order to obtain the benefits of the Act of Congress passed June 7, 1832;

That he enlisted on Turtle Creek, Westmoreland County, state of Pennsylvania, about August or September 1776 for the term of three years, with Colonel Enos McCoy, Major Vernon, and under General Daniel Broadhead, we first marched to the Kittanning on Allegheny River, remained there about two months. The regiment consisting of eight companies then took up the line of march, passed through Hannahs Town in Westmoreland County, Bedford, Shippensburg, Carlisle, Lancaster, and Philadelphia, and on our march met the Hessians when General Washington took Trenton, From Philadelphia we went up the Delaware River to Trenton, then to Bonbrook, while we lay there this deponent with a number of others got into a skirmish with the enemy, when the cock of his gun was taken off by ball, his captain being near, he showed his gun to him, and as he turned away, a ball came across the back parts of his head, cut his hat, cut away parts of his hair, and knocked him down, how long he lay he cannot tell, but when he came too, they were all gone, found himself alone, he happened to take the right course and fell in with his company. The ball lodged in the back part of his head, or neck, and afterwards was extracted by an apothecary in the town. It was long before he got well.

States that he was in the massacre at Paoli, was on guard when the enemy came, he fired upon them and turned around, when a bayonet was thrust into his side, and he fell, he was then struck on the head with the butt of a gun twice. He heard them say he was dead enough, then struck him the third time, and left him for dead. He lay there until the next day and was carried off by them who came to bury the dead. While he lay there next morning and before he was taken off the British Light Horse came along. One of them asked him where the rebels were, he said he did not know, they were about 1,000 strong in number, he told this deponent to come along with him, he said no he was badly wounded, he told him the Devil mend him and left him. This he thinks was the fall of 1777 (best his memory, is greatly failed, especially about dates); States that Captain Montgomery commanded at Bonbrook and then Major Vernon or Vernum took command of the company after Paoli; The company in which deponent was returned west of the mountains and proceeded to Fort Mackintosh, for some time, then went to Fort Lawrence, then to Pittsburg, where he was honorably discharged by Colonel Bayard at the end of his three year service.

Author's note – William Gill's pension application was approved and he received $80.00 per year, for serving as a private for 3 years from the Federal Government. He also received $40 per year from the state of Pennsylvania.

There is no date of death in his pension folder. But his last date of pension payment is the 4th of March 1833.

The first battle he describes, when he has a musket ball lodged in his head/neck is that of Bond Rock on April 13th 1777.

The Battle of Paoli on September 20th 1777 is the next battle he describes where he is left for dead. It is also referred to, as he does, as the Paoli Massacre. This reference is because there were claims that the British took no prisoners and granted no quarter. But William Gill's testimony seems to contradict this with his interaction with the British Light Horse.

Michael Graham

This is one of those extremely well told pensions that details Michael Grahams experience at the Battle of Long Island in August of 1776. He tells his story out so nicely I don't even feel the need to front load the reader with any details.

State of Virginia, County of Bedford, On this 24th day of December 1832 personally appeared in open court before David Saunders, Samuel Hancock, Bowker Preston and George Steptoe, the Justices of the County Court of Bedford now sitting Michael Graham resident in said county of Bedford aged about 74 years who being first duly sworn according to Law doth on his oath make the following declaration in order to obtain the benefits of the act of Congress passed the seventh day of June 1832;

That I was born in the state of Pennsylvania, Lancaster County (now Dauphin) Paxton Township about five miles from Harris Ferry on the Susquehanna River now Harrisburg, that I was seventy four years of age the sixth day of April last. There was a family register which I often saw but my father removed to the state of North Carolina and took it with him and I cannot now say wat has become of it. At the time of my birth my parents were forted for fear of the Indians and a circumstance took place at the time which my mother often related to me that made such a high impression on my mind that I shall never forget the day nor the hour of the day on which she said I was born.

Having served nine months in the Revolutionary War, one tour of six months and another for three as a private and not knowing any person living by whom I can prove it; and having no documentary evidence on the subject do make the following declaration.

That about the last of May or first of June 1776 being then a little turned of eighteen years of age, I turned out a volunteer in a company commanded by Captain J Collier (pronounced Colyer) we composed a part of the Coyles denominated the Flying Camp. We marched by Lancaster, Philadelphia, Trenton, Princeton, Brunswick, Amboy, Elizabethtown, and Newark, to New York where we joined the main army, after continuing some time at New York we were taken to Long Island, and stationed to Brooklyn – There is a high ridge running from the narrows across the island: Below this ridge the British army lay and the Americans above. There is a road leading across this ridge at Flatbush to Brooklyn, the day before the battle eight men were taken from the company to which I belonged on Pickett Guard and posted near the pass and I was one of that number. On the next morning the battle commenced about the break of day or perhaps a little before, at the narrows where Lord Sterling commanded there was a pretty heavy cannonading, kept up and occasionally the firing of small arms, and from the sound appeared to be moving slowly towards Brooklyn. This continued for hours. Although the firing commenced above us and kept spreading until it became general almost in every direction. We continued at our post until I think about 12 o'clock when an officer came and told us to make our escape for we were surrounded. We immediately retreated towards our camp. We had went but a small distance before we saw the enemy paraded in the road before us, we turned to the left and posted ourselves behind a stone fence, from the movements of the enemy we had soon to move from this position, here we got parted and I neither saw officers or men belonging to our party (with the exception of one man) during the balance of that day. I had went but a small distance before I came to a party of our men making a bold stand. I stopped and took one fire at the enemy but they came on with such rapidity that I retreated back into the woods, here I met Colonel Miles a regular officer from Pennsylvania, and Lieutenant Sloan a full cousin of my own, as soon as I had loaded my gun I left them (Colonel Miles was taken prisoner and Lieutenant Sloan killed) as the firing had ceased where I had retreated from, I returned to near the same place. I had not been at this place I think more than one minute before the British came in a different direction from where they were when I retreated, firing platoons as they marched. I turned and took one fire at them and then made my escape as fast as I could, by this time our troops were routed in every direction. It is impossible for me to describe the confusion and horror of the scene that ensued. The artillery flying with the chains over the horses backs, our men running in almost every direction, and run which way they would they were almost sure to meet the British or Hessians and the enemy haggling when they took *prisoners, made it timely a day of distress to the Americans – I escaped by getting behind the British that had been engaged with Lord Sterling and entered a swamp or marsh through which a great many of our men were retreating, some of the were maimed and crying to their fellows to God sake help them out but every man was intent on his own safety and no assistance was rendered. At the side of the marsh there was a pond which I took to be a well, found numbers as they came to this pond jumped in and some were drowned, soon after I entered the marsh a cannonade commenced from our battery on the British and they retreated and I got safely into camp, out of the eight men that were taken from the company to which I belonged the day before the battle on guard, I only escaped, the others were either killed or taken prisoners – At the time I could not account for how it was that our troops were so completely surrounded, but have since understood there was another road across the ridge several miles above Flatbush that was left unsecured by our troops, here the British passed and got between them and Brooklyn, unobserved, this accounts for the disaster of that day. The night after the battle as well as recollect a heavy rain commenced and rained nearly all the time we were on the island, we were at length marched off in the night behind fires made along our entrenchments to the East River, here boats were ready to receive us and we were landed in New York a little before the break of day. We continued but a short time in New York, we were then marched in the night over Kings Bridge and encamped several miles to the east or northeast of this bridge. We continued here I think about two months and afterwards while we lay at this place nothing important transpired in which we were engaged except some little skirmishing with the Hessians – At length the British* made there appearance near a place called West Chester, her some commandeering and skirmishing took place, we then retreated to White Plains, with the main army, here we halted, threw up entrenchments, and waited the approach of the enemy, here an engagement took place between part our troops and the British and every thing seemed to indicate a general engagement, but the British declined attacking us, from this place we retreated a few miles further up the country, here we encamped some time. At length we were taken over the North River and encamped at a village called Spangtown. While we lay at this place, Fort Washington was taken, we could distinctly hear the firing, from Spangtown we commenced our retreat through the Jerseys, passing through Elizabethtown to Brunswick, soon after we crossed the Raritan the British came in sight, here some skirmishing and commandeering took place but it is believed without effect on either side. This was the last sight I saw of the British that campaign. From Brunswick we marched through Princeton to Trenton. this as well as I recollect was about the first of December, our time had now expired we were discharged and I returned home. This was a tour of six months.

During my absence my father sold his plantation with an intention of removing to the state of North Carolina and my oldest brother having removed to Virginia and settled in Rockbridge County and was rector of Liberty Hall Academy. In the spring of the year 1777 I came to Virginia and lived several years with him – In the month of July as well as I recollect in 1781 being then a resident of Rockbridge County I was drafted as a militia man and went under the command of Captain James Gilmour the sub-lieutenants were Samuel McCambell *and John Kelpatrick, our field officers were Colonel Samuel Lewis and Major Long – we marched by the way of Richmond and after many marches and counter marches Cornwallis at length posted himself in New York, from this time to the arrival of General Washington we encamped believed Williamsburg and York. Soon after his arrival we were marched down to the investment of York and encamped below the town, here we continued about four or five days before the surrender of Cornwallis – our time had now expired and we were discharged.*

Author's note – Michael Graham's pension application was approved and he received $30.00 per year, as a private.

He passed away 18 May 1834.

Jacob Grist

Jacob Grist's pension application covers a lot of points I have previously mentioned, and also documents Pennsylvanian's fighting in the southern campaign of the war. He starts out his military service at only 16 years of age, when he substitutes for his father who is called up, or drafted. He actually stands in for his father twice before joining on his own when he turned 18, and the legal age to enter the service. As I mentioned earlier, you could be under 18 to substitute for a family member, but had to be 18 to join or be drafted.

As you'll read in his application, Jacob mentions that when a two-month tour he was on came to an end the reliving militia didn't show up, so he and his company stayed for another tour. This act of a relieving militia not showing up became more frequent as the war drug on and militias had a hard time mustering enough men to march. And usually, the militia whose tour was up didn't hang around for their relief to arrive, or as in Jacob's case fill in for them. The norm was for them to just pack up and go home.

What also interested me about this application was the fact that he details, as a Pennsylvania Line soldier, the fighting he did in the southern campaign of the war. His perspective is very interesting and shows the logistics available to the Continental Army at the time.

State of Pennsylvania, Westmoreland County, On this thirteenth day of April 1835 personally appeared in open court, before the Judges of the Court of Common Pleas of Westmoreland County now sitting Jacob Grist of Hempfield Township in said county aged 72 years, who being duly sworn according to Law, doth make the following declaration in order to obtain the benefit of the act of congress passed 7th June 1832 –

That he entered the service of the United States in the month of August 1779 as a substitute for his father who was drafted in the militia of Lancaster County Pennsylvania, where he then resided. I was enrolled in Captain Paxton's company 22nd Regiment of Pennsylvania militia commanded by Colonel Thompson. I joined the company at Lancaster and was marched to Brandy Wine where we encamped at Gulf Mill, the company to which I belonged was called out on a tour of two months only, when the two months had expired the troops who were to occupy the station had not arrived and we volunteered our service for one month, when we were relieved by troop from Virginia – At the termination of the three months I was discharged. I received a written discharge which I have long since lost – again in the month of June 1780 I served a tour of two months as a substitute for my father (Simon Grist) who was drafted in a company commanded by Captain Gourley in the same regiment of militia above. We marched from Strausburg *to Downingstown from thence to the banks of the Delaware and we encamped at a place called Corrells Ferry. I was discharged about 8 miles above this place and received a written discharge which I have since lost.*

In the month of July (as near as I can recollect) in 1781, I enlisted in the 4th Pennsylvania Regiment then stationed at Carlisle and commanded by Colonel Graig. I was enrolled in a company commanded by Captain Gourley, the same that I had served under in the militia but who had then obtained a commission in the above regiment of Continental Troops and enlisted for the term of eighteen months as a private. We were marched from Carlisle to Little York, where we were kept for some short time under drill. We then marched to 4 Mile Run and from thence to Philadelphia we here embarked aboard vessels lying at Philadelphia and sailed up Christeen Creek to Christeen Bridge, we landed here and marched to the Head of Elk in the state of Maryland, where we embarked and sailed to Baltimore, at Baltimore we went aboard several small galleys and sailed under convoy of the Washington Privateer to the mouth of the James River in Virginia. We passed the French Fleet which was lying off York Harbor blockading Cornwallis' Army at Yorktown, we were landed at the mouth of the James River and soon after joined General Washington's Army, which was lying before Yorktown. After the Battle at Yorktown and capture of Cornwallis and about the latter part of the month of October, the regiment to which I belonged in company with the Pennsylvania Troops under the command of General St. Clair were marched to South Carolina where we joined General Greens Army sometime in the winter at a place called Poupon and were soon afterwards encamped at a place *called Cypress Hill. We had some fighting near this place on the banks at a river (its name I cannot recollect) but think it was called Combea or Combahee. The British had some small craft up the river and landed some troops with whom we had a pretty smart engagement for a short time, we were obligated to retreat across a swamp with considerable loss, but we soon met and succoured by reinforcement of troops under the command of General Guess and the British took to their boats and descended the river. Colonel Lawrence who had the command of the American troops in advance and which I was, was killed early in the engagement. Afterwards I think in the month of March we had a slight skirmish with the British troops on the road to Charleston. The British retreated and made to their shipping.*

Sometime in the month of November or December following General Green gave orders that all those who had served out their 18 months should be marched home. We were marched to Lancaster in Pennsylvania and after lying there some time we were discharged by Colonel Hampton. I received a written discharge which I kept for a long time afterwards in my trunk, but it has been lost now for many years and I am unable to tell what became of it. As well as I am able to remember it was about twenty months or more from the time I was enrolled at Carlisle until I was discharged at Lancaster.

Author's note – Jacob Grist's pension application was initially denied by the Pension Office. But it was resubmitted for appeal. The appeal was partially approved in February 1838. The first part of his service was denied because it could not be verified. He was awarded credit for the 18 months of his service due to the fact that he provided a living witness who served with him to testify to his service. There is no mention as to how much he was granted as a pension. There is also no date of death given.

He mentions fighting at Combahee, which would be the Battle of Combahee that took place August 27th 1782, in which Lieutenant Colonel John Laurens was killed.

James Guffey

In another really interesting account of the frontier fighting all the way to the west of Pennsylvania, James Guffey describes a lengthy service around Pittsburgh and Fort Pitt. In his early service he is predominantly scouting, spying on the Indians, and most importantly carrying correspondence. During this time that is the only means of communicating. There is no telegraph yet so if you want to deliver a message or coordinate a military event with someone you need to do it through hand written correspondence. And the only way to have that delivered was by having someone hand deliver it. And this was by no means an easy task. As you'll read it was very dangerous traveling from one outpost to another in basically unmarked hostile territory.

In looking at his monetary pension breakdown I also thought it was a good time to sort of explain the odd sums in annual pension payments you have been seeing. You'll see at the end of this chapter that Mr. Guffey is granted an annual pension of $136.66. This is derived from the fact that he is granted one month credit for serving as a private and awarded $3.33 per year for that service. He is also granted 10 months credit for serving as a lieutenant and awarded $133.33 per year for that service. So, his total annual pension comes out to $136.66. Unfortunately, all of his service up to the point of becoming a private is not credited. This is probably due to the fact that it was from around 1773 to 1775 and pre-war service.

State of Indiana, County of Jackson, on this thirteenth day of May in the year 1834, personally appeared in open court before the Honorable Abel Findley Judge of the same being the Probate Court of record of the same county of Jackson and state aforesaid now sitting at the court house there of in Brownstown, James Guffey a resident of Hamilton Township in the same county and state, aged about 76 years on the fifth day of May 1834, who being first duly sworn according to law doth on his oath make the following declaration in order to obtain the benefit of the Act passed June 7, 1832.

That he entered the service of the United States under the following named officers and served as herein stated that he Applicant was born the fifth day of May in the year 1758 in what was called Nods – Forrest, now Harford or Hartford County in the state of Maryland where he remained until about ten years or more of age, his father being dead, the Applicant was removed by is mother across the Susquehanna River (thinks into Cecil County) in the same state where he lived with his mother about two years or upwards and afterwards he removed with his mother to Pittsburgh, now Allegheny County in the state of Pennsylvania being about one year on the route to that place having stopped during the winter at Kenocucheay, when Applicant arrived at Pittsburgh the Fort there was as nearly as he can recollect commanded by General Hand, Applicant recollects that he arrived at the Fort the September or sometime in the fall following and that there was a treaty with the Indians, or some of them pretended to be made, about that time and place Applicant arrived in Pittsburgh. Applicant though very young, not being more than about fourteen or fifteen years of age volunteered and entered the service to and assist in keeping the fort at that place and defending the frontier in that vicinity about two years. Applicant especially belonged to the service in and about the fort during that time. Applicant does not now recollect to what company or regiment he was particularly attached during that time as he was taken (immediately on him entering the service) under the immediate direction and assistance of General Hand, then commanding the fort at Pittsburgh and Applicant was often employed to assist an Indian spy during that time to bring in the Americans that were often and repeatedly killed many times by the enemy during that time to the fort and to bury the dead and assist in taking care of the wounded and to carry expresses, especially for General Hand while he commanded the fort then. And a year after Applicant entered the service there as aforesaid, or at his direction, during which time Applicant often went on express under the direction of General Hand. Applicant recollects that on one occasion he was unwilling to go by himself when directed by General Hand to carry an express to Wheeling, and that as nearly as Applicant recalls no one was willing to go with Applicant, and Captain William Flinn who at the request of General Hand went with Applicant; the unwillingness of any person to go was owing to the route being beset as was supposed by the enemy who were in the neighborhood killing the American inhabitants and destroying *their property. Applicant and William Flinn on that occasion performed the perilous duty of carrying the express. On another occasion Applicant alone by request of General Hand carried an express to Logstown, a station below Pittsburgh. Applicant often carried expresses which were dangerous. Applicant mentions the two above in particular in being very dangerous, the events having made a lasting impression on his memory. Applicant recollects that in carrying the express to Wheeling as above mentioned he and Flinn went on the opposite side of the river from Pittsburgh. Applicant also recollects that it was the second season after he arrived at Pittsburgh from Cecil County Maryland when he entered the service in the fort as above stated, he things General Hand commanded the fort about one year after applicant entered it and that Colonel Neville afterwards commanded awhile and afterwards Captain or Colonel Heith, does not recollect distinctly the title, Colonel Gibson also commanded there awhile but Applicant does not recollect when he took command. Captain or Colonel Heith commanded the fort about the time that Applicant left the garrison at the expiration of his two years service, during which time he was on many scouts after and spying the Indians, carrying expresses, and at the end of his service above stated Applicant with many others who had volunteered to defend the fort and frontier settlements left the fort about the same time, and that shortly afterwards Applicant enrolled himself in Captain Rollator's company in the regiment then commanded by Colonel Stinson of Pennsylvania militia, now Allegheny County, when Applicant was about seventeen or eighteen years of age in the spring of the year, and that during the summer following Applicant does not recollect the precise time of summer he was drafted to go and serve a tour of about one month or upward under Captain Rollator to build Fort Crawford, which Applicant assisted in building during that time on the water of the Allegheny above Pucketty Creek, but during the most of that time Applicant was engaged as an Indian spy with James King, the duties of which he assisted King to perform.*

Applicant was then young but chosen by King who was older to assist him in the duties. Applicant recollects Colonel Crawford with whom he was well acquainted and recollects Lieutenant Morgan and he thinks Parks, who he understood to be of the regular service, when then Fort Crawford was finished Applicant was discharged but received no written discharge, was nearly about eighteen years old, does not recollect the date precisely. At that time about two weeks after applicant returned home he received an ensign commission which was sent him be believes or suspected by or through Colonel Crawford, but does not recollect the name or inscription of the regiment in the Pennsylvania troops, and Applicant held his commission about two years in the same company of militia commanded by Captain Rollator as above stated, during which time Applicant was repeatedly and very often in dangerous and perilous service in short tour against the enemy defending the settlements there.

Applicant afterwards received a lieutenants commission in the same company of Pennsylvania militia, which was afterwards commanded by Captain John Turner who succeeded Captain Rollator who had resigned, all of which or nearly as Applicant recollects took place in or about the month of March of that year and the same spring the militia being all classes, and put under strict regulation, , a portion of them in continuous service, each class was to a definite period of time according to orders as required by the commanding officer, and Captain John Turner refusing to command or to go into service, Applicant was assigned the company or portion to take the place of Captain Turner, and also to command another company of militia immediately in succession, each a tour or campaign of ten days, making together twenty days at a time, which service of twenty day tours Applicant performed every thirty days out of every fifty days for two years. That during the last mentioned service he was regularly the rank of and command generally of lieutenant, though he often acted or had the command of captain, which service was rendered principally in the neighborhood of Big Pleasant Creek and Little Pleasant Creek guarding and defending a district of the frontier settlement there and some times as far as Fort Crawford, that during the remainder of the time between the twenty day tours above mentioned Applicant was often engaged as an Indian spy or scouting to protect some part of the frontier or training and preparing the militia in or about to go into a tour or campaign of service, so that during the whole of that two years last mentioned Applicant was employed exclusively in the service of his country during the Revolutionary War, that during these two years the Applicant was out in actual service commanding and wearing the rank of lieutenant as aforesaid about as nearly as he recollects and believes sixteen times or tours, twenty days such as above stated, which are equal to about ten months actual service as lieutenant, in addition to his month service as a private while building Fort Crawford as above stated.

Author's note – James Guffey's pension application was approved and he received $136.33 per year, as a private and lieutenant.

He lived almost three more years after giving his deposition to the court and passed away on the 25th of December 1837.

James Guthrie

It's interesting reading James Guthrie's application because he seems, like a lot of men of his time, to want to keep moving west towards the frontier. Some of these pension applications you read are filed in Kentucky, which would not become a state until 1792. That's almost 10 years after the Revolutionary War. During this time, it was wild and dangerous Indian country. It's also interesting to note that his deposition is taken in his home, which is very unusual. He was in fact too ill to travel the 12 miles to the court house where they are typically taken. One more thing to note is that he had one of the biggest pension folders I have come across, at 128 pages. Most of the ones you will read in this book have 10 to 20. With all these documents I found several later versions of his deposition as he clarified details needed to process his claim. I have made notes where there was clarifying information from his original statement.

State of Kentucky, Jefferson County, On this twenty third day of April one thousand eight hundred and thirty three, personally *appeared before the undersigned James Pomeroy a Justice of the Peace in and for the county of Jefferson and state of Kentucky and by virtue of his commission one of the Judges of the County Court of said county, James Guthrie (at his own house) a resident in said county and state aged eighty three years on the 25th day of February last past, born in the year 1749 or 1750 O.S. (as it was then expressed) who being first duly sworn according to Law doth on his oath make the following declaration in order to obtain the benefit of the provisions made by Act of Congress passed June 7th 1832 sowit, that he was born in the now state of Delaware, New Castle county and remained there until the year 1773 and then moved to Bedford County in the back part of Pennsylvania, that in 1776 the court or committee of Bedford County ordered three companies to be raised and some time in June of 1776, say the 15th, this applicant entered the service of the United States, state of Pennsylvania as he then thought for three months; the companies were commanded by Captains Swearingin, Carson, and Miller, and the present applicant was lieutenant under Captain Miller, that some time in July 1776 orders were given to raise a regiment in Bedford County of Pennsylvania to be known by the name of the Eighth Pennsylvania Regiment and the said regiment was raised and was commanded by Enos McCoy – Colonel, George Wilson – Lieutenant Colonel, Richard Butler – Major (which last was killed in St. Clears defeat by the Indians) and that Captain Miller commanded the company in which the present applicant belonged and in which he was lieutenant.*

That when the regiment was organized we were stationed on the Allegany River until sometime in December 1776, we then were ordered to march and passing through the towns of Ligonier, Bedford, Shippensburg, Carlisle, Lancaster, and Philadelphia, there crossed the river and passed through Trenton, Princeton in New Jersey, and the Basken Ridge joined the army at Quibbletown, we had several little skirmishes and near Bound Brook we were defeated and lost our artillery, we were then under the command of General Lincoln;

Author's note – He adds a date of 13th of April for this engagement.

we then went to Morristown New Jersey then to Goshen on the highlands, York (now je believes the state of New York) then to Germantown Pennsylvania. At Trenton Colonel McCoy died and at Quibbletown our lieutenant colonel died and we were afterwards commanded by Colonel Daniel Broadhead. At Germantown (after having served fully fourteen months as lieutenant this applicant in August 1777, say after the 15th of same month, he resigned his commission to Colonel Broadhead. This applicant states that he had a commission but does not now recollect by whom it was signed, that he gave it to Colonel Broadhead, and that he received a discharge in writing from Major Byard, but does not now know what has become of it. This applicant states that there are now no persons living by whom he can prove his time he entered the service and the time he left, but that he can prove his service in part by James Welch, and that he served as a lieutenant in the Eighth Pennsylvania Regiment.

After this applicant left the regular service he went to Cumberland County Pennsylvania and was enrolled in the militia of said county in Colonel Culbertson's regiment and in the last of August 1778 he was called under the then laws of Pennsylvania to perform a tour of two months which service he rendered as a private soldier under Captain Jack and was stationed on the Juniata River in Canoe Valley as a guard to the frontier settlement and that he left the service in October 1778 after his two months expired. This tour he can prove in part by the said James Welch but by no other person.

This applicant states that in the year 1779 he moved to the now state of Kentucky and landed at the Falls of Ohio (Louisville) on the 6th day of April said year, and on the 13th day of May 1779 this applicant joined a volunteer company under Captain William Harrod. The party was commanded by Major Bowman; the object was to fight the Indians who were destroying the frontier settlements. That they marched through the Indian country to one of their towns on the head of the Little Miami River, the town was called New Chillicotha and had some fighting with the Indians and applicant was wounded in his face,

Author's note – In another claim he states that he was severely wounded and the engagement was known as Bowmans Defeat and in another The Battle of Piqua Chillicotha.

and on his return to Kentucky he continued in the service until some time in November (he thinks sometime about the middle) of the same year, employed sometimes as a spy, a hunter, and in the fort at Lexington Kentucky. This tour he cannot prove by any living witness that he knows except in part by James Welch. He served this tour six months.

And in the year 1780 on the 20th or 22nd of July he volunteered to go a campaign against the Indians and joined a company *commanded by Daniel Hall, this campaign was commanded by General George B. Clark, Colonel William Lynn, Lieutenant Harrod, and Major Edward Bulger. That they marched to the waters of Little Miami, now state of Ohio, and onto the Pickaway Towns on the waters of the Big Miami, called Mad River, and there had some fighting with the Indians and destroyed several of their towns and that the party returned home the last of August of the same year, that he served at least thirty days at this time and can prove his service by James Welch, but by no other person that he knows of living.*

That in the year 1781 this applicant was called upon to aid in building a fort at the Falls of Ohio and in constructing a boat to sail between the falls and the mouth of Kentucky River in order to guard off the Indians against persons moving to Kentucky, and that he spent thirty days in this service, but he knows of no person by who he can prove his service.

That in the year 1782 this applicant volunteered again under General Clark to go against the Indians and belonged to a company commanded by Captain Samuel Pottenger, that he marched to the west branch of the Big Miami and destroyed several Indian towns and had an engagement with the Indians at a placed called High Rock and that he was in the service forty days in the month of, can't recollect, which tour he can prove by William Tyler.

This applicant states further that from the time he moved to Kentucky in 1779 to the time 1791 he was almost a soldier the whole time, that he spent his time in the different forts and garrisons in said state and was almost constantly employed as a soldier in defending the country or spying the movements of the Indians.

Author's note – James Guthrie's pension application was approved and he received $173.33 per year for 12 months service as a lieutenant and 4 months as a private.

He passed away on the 24th of march 1841.

Major Richard Butler went on to become a Major General and was killed at St. Clair's defeat in November of 1791.

Colonel Aeneas Mackay (referred to here as Enos McCoy) passed away at Quibbletown sometime in February along with 50 other men from cold related illness[6]. Lieutenant Colonel George Wilson died the night of 24/25 February 1777 of inflammation of the lungs.[7]

Barnet Hageman

This is a relatively short story but I thought his description of being wounded in the southern campaign was very interesting, along with his fighting at Brandywine.

State of Ohio, Wayne County. On this 18th day of March A.D. 1833 personally appeared before the Judges of the Court of Common Pleas of Wayne County Ohio, Barnet Hageman, a resident of Plain Township in the county of Wayne and state of Ohio aged seventy seven who being duly sworn according to Law doth on his oath make the following declaration in order to obtain the benefit of the provisions made by the Act of Congress passed June the 7th 1832.

That he enlisted in the army of the United States in the year 1776 or 1777 as near as he can now recollect with Colonel Butler and served in the Second Regiment of the Pennsylvania Line on the continental establishment under the following named officers, Colonel Butler commanded the regiment and Captain Davis commanded the company, that he enlisted for the term of three years and served to the expiration of the term of his enlistment in the county of Bucks in the state of Pennsylvania – That after he joined his regiment he was stationed at a place called The Trapp about thirty miles from Philadelphia where he was in a skirmish with British troops. He was from thence marched with his company to Newport in the state of Delaware where he remained for four or five weeks and from thence he was marched to the state of South Carolina. That whilst this deponent was stationed at the Horse Plains in South Carolina he was ordered on a scouting party against the enemy and was surrounded by a party of British Light Horse and in the affray this deponent received a blow from the sword of one of the enemy across his head, which felled him to the ground and in the attempt to ward of the blow he nearly lost the fore finger of his left hand which was almost cut off. After this deponent had recovered from the effects of the blow, he found that the enemy had left the ground and were perusing the Americans in various directions. He then made for a small thicket where he remained, concealed for some time, when one of the British horseman returned near where he lay he shot the horseman and succeeded in getting possession of the horse and equipment, and brought them safe to camp, where he gave them up t the American commander. He was then marched back to the state of Pennsylvania after an absence of about two years. This deponent then joined the main army and was in the battle of Brandywine from daylight until dark, and was wounded in the foot by a musket shot. The next night after the battle the troops fell back upon the highlands and built large fires and after the fires were kindled they left the ground and were marched to Wilmington and from thence to Germantown where they where again engaged with the enemy in the Battle of Germantown. When this deponents term *of three years service having expired, he was discharged from the continental service. This deponent believes that his discharged was signed by George Washington. He retained his discharge until about three years since when his chest in which he kept his papers was broke open and his discharge taken away or destroyed.*

Author's note – Barnet Hageman's pension application was disapproved due to his lack of proof of service. There was also no date of death given.

Frederick Hain

In this application from Frederick Hain there is an interesting display of patriotism from his father, which drew me to his application. He also mentions some interesting facts, like the amount of bounty he received for enlisting, an obscure law that allowed his brother and himself to get out of service by recruiting another able bodied man, and his explanation for him seeking a pension. I also found several of his other letters concerning his claim interesting and have made comment of them in my author's note at the end of this chapter.

State of Pennsylvania, County of Berks, on this fifth day of September Anno Domini on thousand eight hundred and thirty four personally appeared in open court, before Garrick Mallery esquire, President and his appointed judges of the Court of Common Pleas for the County of Berks, aforesaid, now sitting Frederick Hain, a resident of the township of Heidelberg in the above county of Berks and state of Pennsylvania, aged eighty four years, who being first duly sworn according to Law, doth on his oath, make the following declaration in order to obtain the benefit of the Act of Congress passed June 7th 1832;

That he entered the service of the United States under the following circumstances, and under the following named officers and served as herein stated;

That he was born on the 15th day of September Anno Domini one thousand seven hundred and fifty in the township of Heidelberg aforesaid and county of Berks – That the record of his age is in the family bible as well in the Archives, on record of "Hain's Church" in the said township of Heidelberg – That he resided in the place of his nativity when the Revolutionary War broke forth, and after independence being proclaimed and peace again restored he returned to his native residence in the above mentioned township and country – That in the month of December of the year 1776 he took a bounty of fifty shillings to the best of his recollection and knowledge from Valentine Eckert then a resident and chief Burgess of the now borough of Reading in Bucks County aforesaid and then became enrolled in Captain Fishers company of aforesaid and then became enrolled in Captain Fishers company in the 1st Regiment of the Pennsylvania Line commanded by Colonel Broadhead – That he remained only a few days in Reading when he marched in company with others to join his regiment, then stationed in Philadelphia, or near it – That his company was ordered to Trenton and that he was one of the guards who conducted the Hessian prisoners to the aforesaid borough of Reading – That he was a sergeant from the very day he entered and performed the duties of that office – That he in company with a man named Mathias Wenrich got orders to recruit in the borough of Reading aforesaid and neighborhood, that accordingly they did *do so and enlisted many at the Sinking Springs, throughout the country, and in the borough of Reading itself. That according to the resolution and recommendation of Congress made the 14th of April A.D. 1777 respecting substitutes, that himself and his brother Daniel Hain found and caused to enlist during the war an able bodied man, by the name of John Sutton on the 5th day of May 1777 in the company of Captain George Ross of the 11th Pennsylvania Line*

Author's note – This law basically stated that Fredrick and Daniel Haines didn't have to serve in the military as long as John Sutton was serving, since they recruited him.

That he returned home and attended to the business of his mill (being a miller by occupation) that his father Henry Hain was a true republican, and almost a fanatic in the cause of freedom, and that on his father hearing of Arnold's treason in the fall of 1780, he insisted on all his boys, namely Adam, Frederick, Daniel, Joseph, and John to go to the war, and applicant stated that his father himself went along also and on his appearing before Colonel Broadhead, the Colonel said it was a shame to suffer so old a man to perform the arduous duties of a soldier, and much against fathers will, prevailed on him to return home again. That Adam Hain and Joseph Hain, above named, were captains to the best of applicant's knowledge and recollection, that he, applicant, was then made ensign in Captain Fishers company – aforesaid and served to the beginning of June 1783 – That he gave Valentine Eckert $700 for the purpose of enlistment, the same who was chief Burgess in the borough of Reading aforesaid and never received a cent since of it –

Author's note – After reading another letter he wrote mentioning this payment it seems to be a loan or donation for helping the recruitment cause.

That his team of horses and wagon and servants were every week during emergencies performed some kind offerings or other cause of liberty and our country's government – That he would not even now make any application of the present kind, but that he sees others that done less for the country and cause of liberty than he did receiving large pensions, and otherwise better off in the world than he is – That in as much as he never expected any recompence for his service. That he was carless of his papers – That he has his discharge, dated 19 June 1783 signed William Huston Adjutant – He believing of General Lincoln's division, but otherwise deciphering that is unintelligible, his commission, he long since lost with his pocket book on a hunting occasion. He knows of no human being who could prove his service now alive, the only one in his neighborhood was Stephen Hossier who served under him, but unfortunately is dead now almost two years.

Author's note – In another deposition in which he is clarifying some information he also makes this statement on his service; *"That he narrowly escaped with his life from Paoli, and that Colonel Broadhead was taken prisoner there by a soldier who appeared to him to be a Hessian, that the colonel offered him his watch to let him go, that the soldier accepted of it and let him escape – cursing him at the same time in the German language."*

Author's note – Frederick Hain's application was denied because the Pension Board said "*He did not serve under military authority."* He did write several passionate letters appealing his case and in one of his final letters he states that he is frustrated with the process, which is wanting him to basically start his application process from scratch. He goes on to say that he is basically out of time and *"I am now nearly 85 years of age and the cold finger of death beckoning me to the tomb."*

In another letter to the Pension Board, he closes with *"If in your judgement you think proper to award me nothing I must submit, I say with painful reflection that I did more for my country in the hour of peril than any man I know of within 20 miles of my residence."*

There is no indication in his folder to when he passed away.

John Hall

We finally come to our first mariner story! Being a retired seafarer, I really enjoy coming across these rare claims. And his story is very similar to the legendary Captain John Paul Jones and his ships the Bonhomme Richard, or in English the Good Man Richard (Dick) and Ariel - although his time line and events don't match up with those of the legendary captain and his ships. But then again, he is trying to remember back 71 years. And I find that these veterans usually start their stories with what's the most vivid in their memory, not what really happened first.

I'll give you some background information on Captain Jones and his ships so you will have some context, and I think you'll see that John Hall is pretty close in his story. The Bonhomme Richard was placed by the French at the disposal of Captain John Paul Jones in February 1779. But she was sunk by HMS Serapis in September 1779 off the coast of England. As the two ships were slugging it out and the Bonhomme Richard was heavily damaged, the captain of Serapis called for Captain

Jones to surrender. In reply he uttered his famous response "Sir, I have not yet begun to fight."

Ariel was lent to the Continental Navy by the French from October 1780 to June 1781 and commanded by Captain John Paul Jones. She sailed from France with much needed gunpowder and other military supplies and reached Philadelphia in February 1781. Along the way she engaged the British privateer Triumph and subdued her with a volley from her guns, but she escaped. In June 1781 Ariel was turned over to the French in Philadelphia and sailed back to France. She was scuttled in 1783 on a river in Belgium.

State of Pennsylvania, Berks County, on the 4th day of October 1850 personally appearing before me a justice of the peace in and for the county of Berks and state of Pennsylvania, John Hall, a resident of Leesport County and state aforesaid aged eighty nine years, who being first sworn according to law doth on his oath make the following declaration in order to obtain the benefit of the act of congress passed June 7th 1832;

That he went on sea, the service of the United States under the following named officers as herein stated; went to sea in the year 1779 in the month of August on ship Ariel, I entered under Captain Courder at Philadelphia in the state of Pennsylvania as a sailor. The ship Ariel was loaded with some tobacco at Philadelphia, which we have unloaded at Lorient in old France. And at the same place we was dispatched under Paul Jones, captain, an the ship by the name Good Man Dick, with a load of powter to take to Philadelphia for ammunition for the war. Our first lieutenants name was Herman Diehl and our second lieutenants name was Benjaman Lunds, sailor master Herman Stacy.

One morning on sea we was attackted by on English ship, the name of it I do not know. And the next day again we give them four brought sides and we received two brought sides. They have attackted us the second time. Fires was given on both sides, our men worked like brave soldiers and we have shed their masts off and took about thirty of them prisoner. We have took them along to Philadelphia, there they was locked up and what became with them afterwards I do not know, and the powter we loaded on our ship we unloaded at Philadelphia and took it to the magazine. The ship Good Man Dick belonged to the King of France, he has sent afterwards a fricket and men to take the ship Good Man Dick to France, and between Philadelphia and New York the fricket and ship was both taken by the English, our captain, Paul Jones was a good man; in did and further do depose and say that he was on sea about three years and that Captain Paul Jones give him his honorable discharge at Philadelphia in the state of Pennsylvania where he was called into the service.

Author's note - Unfortunately his folder is very small and only states that his claim was rejected. There is no reason for the rejection and no date of death given. And like a true sailor he signs his deposition with his mark, an "X."

Joseph Harborn

Here is another very interesting nautical story. Although his file was very short, it did document his deposition, a pension rejection notice, a letter from his lawyer asking for a status of his claim, and a letter from Stephen Decatur vouching that Joseph Harborn served under him.

The Stephen Decatur that he served under was the senior, and not his more famous son who took his father's name and captained many famous ships to include the USS Constitution.

This claim was written very roughly and pretty much as Joseph Harborn dictated it in his salty seafaring dialect, which I think adds to the uniqueness of the story, but may require additional author's notes.

Philadelphia Nov 6th 1824. Your portioner is an old revolutioner born in Philadelphia 1749, Dec 27; I sailt on board Polly Cape Long bound to St. Croix the year 1774, nothing particular happened that voyage, after this I sailed again in the same vessel bound to St. Croix. The third voyage in said vessel I was taken by Captain John Phillips in sloop Revenge of Antigua and coming through Sail Rock passage was retaken by Captain Waters in the Baltimore Hero mounted 10 guns, he set me ashore at Depais, from then I got up to Bastare,

Author's note – Dieppe and Basseterre are towns on the island of St. Kitts.

then I got down to St. Croix, and I met Captain John Beans in sloop Liberty with 10 guns who wanted me to enter, but I being apprehensive that he would keep me some length of time I signed the articles when he promised me upon his honour to give me my discharge at the first port we make on the continent of America. Then I went to work, we took in salt for balance, some rum, some chests of arms, and a quantity of gun powder, I can't tell how much, then I proceeded on our home, and passage we meet two small privateers, we beat them off as our orders was not to chase them, we saw no more them till we got to Bermuda where we fell in with the 74,

Autor's note – A 74 was a British ship of the line having 74 guns.

she was calmed and they hoisted out there boat and lowered one of the quarter deck guns into the boat, then they came to us with the boat full of men and they hove a signal for us to show our colours, our captain would not show them our colours till they fired a gun, then we swept her brood side around to the boat and fired grape shot as far as would reach, and then fired round shot, a breeze started up about the middle of the night and we cleared them in the dark, we were chased from one day to another till we got off Cape Fear to the southern, we got up then to Edenton N.C. I was apprehended to take those of legions to the country, which I complied with, then I travelled by land to Williamsburg Virginia, there I was brought a four Colonel Mason, there to take those of legion again, then I travelled down towards Philadelphia were I remained till the English landed at the Head Of Elk.

Then I went off to Baltimore when the English got into Philadelphia, I remained there as barkeeper with Jacob Myers, tavern keeper, corner of Gay Street, there I married, and when Lord Howe went out of Philadelphia, I came in and brought my wife. I looked for another vessel to get satisfaction, I salt with Captain Harmitage in the year 1779 and went out to sea, the 3rd day we ere out we met with three privateers belonging to Gutrage, we engaged them two hours and a half and there I was taken and carried to New York, there I lay a prisoner, that winter when I saw David Sprout come aboard of Old Jersey as he was commissary of American Prisoners.

Author's note – Old Jersey was the HMS Jersey which was converted into a prison ship.

I begged of him to get me exchanged as soon as possible, in about a week after that there was a Flag Ship come along side the Old Jersey, a boat loaded went ashore with dead every morning to be buried in the sand. Mr. Sprout sent his Flag Ship along side and took 300 of us in which I was among, and I came up to Philadelphia. I went to one Blair McClenachan - a noted merchant to return him thanks for kindness to my family while I was a prisoner, while I was there old Steven Decatur came in and asked me if I was one of the men that got exchanged, I told him I was, he asked me if I would try again, I said I would and went with him aboard the Fair American, I sailt with him and went out with 150 of us aboard and took 13 sail of English vessels and got them all safe in, we broke up her cruise then for the honour of 13 states, and then we went out again – Fell in with the Hulker and cruised together off Charleston Bar where we took some fine prizes, we fell in with the ship Richmond, we engaged her from 12 o'clock at night till 4 o'clock next day in the afternoon when colours came down, we brought the captain of Richmond and all the well men with us to Philadelphia, but we had the misfortune to lose the prize – The honorable board will be so good as to let me have a pension for it to live upon in my old days. I a, now 75 years old.

Author's note – The Hulker was a privateer like the Fair American and the two of them captured HMS Rodney on October 7th 1780 off of Charleston, South Carolina. And on the 14th of October 1780 they captured HMS Richard – who Joseph Harborn is referring to as Richmond.[8]

James Hays

It's not too often you find men shifting from the army to the navy, and back again. That makes James Hay's story unusual. When he joins the army, he is in an artillery company which is a first for this book. He then joins on with the brig Baltimore - an armed convoy or packet vessel. She is listed as a 12-gun vessel, but he refers to her as having 10 guns. With his previous tour in the artillery, I would have guessed that he worked with the guns on this vessel, although he states that he went aboard as a carpenter. After being ship wrecked, he must have had enough of the sea and rejoins an artillery company once again in the army.

State of Pennsylvania, city of Philadelphia. In the Court of Common Pleas in the court of Philadelphia on this twenty fifth day of April Anno Domini 1833 personally appeared James Hays of the city of Philadelphia7 and state of Pennsylvania, aged seventy eight, who being first duly sworn according to Law doth on his oath make the following declaration in order to obtain the benefit of the provision made by the act of Congress passed June 7 1833.

That in the year 1776 he lived in the city of Philadelphia where he had been born. In that year he joined as a volunteer a company of artillery of Pennsylvania troop in the service of the United States commanded by Captain Stiles, the company was raised at Philadelphia – it belonged to the 3rd regiment of state artillery, Colonel Eyre – he marched in the summer of 1776 with his company, he being a private, to Amboy in New Jersey – where he was engaged in camp duty in the neighborhood of the enemy for two months and upwards – they were marched back to Philadelphia and this declarant was honorably discharged from active service, still continuing a member of the company – In the next year, he went with the same company under Captain Fry (who had been first lieutenant at Amboy) down the Delaware to Billingsport New Jersey – they were driven from there after some weeks by the British troops. And went to the Pennsylvania side of the river to Fort Mifflin – after some days they were marched to Head Quarters White Marsh, Montgomery County Pennsylvania north of Philadelphia – they were camped there and remained till about Christmas when they were discharged – The British being in Philadelphia – the declarant went up to Allentown Pennsylvania and worked at making cartridges in the laboratory, then under charge of his old Captain Stiles – From the time he went to Billingsport till he was discharged at White Marsh it was four months at least – during which time he was in military service.

After the British left Philadelphia, he returned to the city and worked for a time at his trade as a carpenter – He then entered *as a carpenter on board of the United States Brig. Baltimore, a packet, - Captain Nicholson, then fitted out at Philadelphia for the purpose of convoying ships with provisions for the French fleet at Cape Francis -*

Author's note – Cape Francis is in Haiti

The Baltimore carried ten, four pounders, she sailed with her convoy of 5 vessels and arrived at Cape Francis with two of the vessels, the other 3 being lost in a gale and never heard of again – after repairing the effects of the gale on the Baltimore, and remaining some time in the West Indies – they returned homewards, and were castaway on Cape Henry in 1781 and lost everything –

Author's note – The official date of the Baltimore sinking off of Cape Henry, Virginia is 29 January 1780.

Declarant after much privation and suffering reached Philadelphia by land – He served the United States in the Baltimore full eight months.

After he came home he joined an artillery company and was called into active service for two months – The company was a volunteer corps and commanded by Captain McCulloch and belonged to Colonel Marsh's regiment of state troops in serving the United States – they were sent down to Billingsport and served in barracks full two months, when they came home and were discharged.

He says that in addition to the services already mentioned he went as a volunteer with Commadore John Barry on an expedition up the Delaware towards Trenton in boats – It was after he had been at Amboy and in cold weather when the

Hessians were at Trenton. He served on this occasion five or six weeks.

Author's note – James Hay's pension application was approved and he received $62.66 per year for 9 months service as a seaman in the navy and 8 months service as a private in the army.

There is no mention of his date of death.

In an interesting affidavit from Jacob Wayne, who was a character witness to James's service, Mr. Wayne states that he witnessed the sinking of the Baltimore Packet and the crew swimming ashore. Mr. Wayne was himself on the vessel Mars and she sank soon after the Baltimore Packet. Just in a side note Mr. Wayne says that the Mars was chased into St. Eustatius, a Caribbean Island, by the British. There they took on cargo and were headed to the Chesapeake Bay. Knowing that the Baltimore was armed, the Mars more than likely fell in with her on her way to the Chesapeake Bay as protection. That would explain why they were traveling together when they were sunk. He ended up on shore with James Hays and the other survivors of the shipwrecked crews. Together they walked to Portsmouth, Virginia.

Jacob Hefflebower

One of the more elite members of the military at this time were known as the Grenadier. Typically, they were associated with the British Army, but there were some in the Continental Army. They were usually chosen by their size and strength and were probably the fittest of the fighting troops. And unlike regular uniforms, they wore a special one which included a tall cap that had a metal plate on the front. I didn't come across to may Grenadiers looking through pension claims, so that in itself made Jacob Hefflebower's story unique. But he was also near the famous General Mercer when he was killed at the Battle of Princeton. And during this battle he was severely wounded in the head.

Because his initial pension claim deposition was augmented by a more detailed clarification letter, I will include that at the end of his initial statement.

State of Pennsylvania, Dauphin County. On this twenty first day of August A.D. 1833 personally appeared in open court in the Court of Common Pleas of Dauphin County aforesaid, preparing a clerk and having a purple seal and being a county record by virtue of the constitution of the Commonwealth – Jacob Heffelbower, resident in Derry Township in the county, aged seventy seven years, who being duly sworn according to Law, doth on his oath make the following declaration in order to obtain the benefit of the provisions of the act of Congress passed June7 1832.

That near the beginning of the Revolutionary War, he marched as a substitute in a company of militia from Conewago in the County of Lancaster in this state"

Author's note – He substituted for his father George Hefflebower.

The said company being commanded by Captain Thomas Robinson – that they marched to the state of Jersey, hence, after remaining about two months, the said company returned home – Deponent was then about 19 years of age – He declined returning and enlisted for the term of three years into a company of United States Grenadiers commanded by Captain Smith – attached, or deponent believes to the third regiment of the army of the United States – Deponent was marched thru Jersey, for a considerable time – he lay some months on an island, which he thinks was Long Island – he also lay some time near a fort, which he thinks was called Red Bank, He was at one time under the command of Colonel Housecker, who deserted to the English from the island expressed by the deponent. Deponent served in the army without interruption for upwards of one year and eleven months previous to the Battle of Princeton – He was with his company in the right wing of the army in the battle, where he received a severe wound in his fore head from a bullet – at the time of his wound, deponent had upon him his Grenadiers cap, the front of which was a brass plate, upon this the bullet struck, and he thus escaped with his life – Deponent fell to the ground and after the battle was carried from the field by John Gundrum, now living in Lebanon County, and his whole statement is certified by esquire Wise of Lebanon County, and is her to attached -Gundrum is now blind – The division of the army to which deponent was attached, separated from General Washington about 10'oclock of the 1st January and marched toward Princeton under the command of General Mercer, and reached Princeton about one and a half hours after sunrise, on the morning of the 3rd of January 1777, and deponent distinctly recollects – as soon as we reached the rising ground near to Princeton, the British soldiers (no Hessians being among them) fired upon us – Having the advantage of the ground, we kept at it, being protected from the aim of the enemy, by the smoke of our own guns, which the wind blew towards them -Being above the enemy, their fire generally fell short of us – We continued engaged a considerable time, neither party advancing, till some field pieces were brought up behind use and being fired upon the enemy, they surrendered – by raising a white flag – General Mercer was engaged about the middle of our division. He parted under his immediate command, was attacked by the enemy with bayonet – Deponent was within 50 or 60 yards of him when he fell, near to the American standard, pierced by bayonets of the enemy, he called aloud, and cut down two of the enemy with his own sword.

Author's note – The British thought they had bayoneted General George Washington.

The afternoon previous to the battle, some snow and sleet fell, but the weather cleared off towards evening – The night was pretty cold – but the morning of the battle, was good weather – The wound which deponent received at the said battle, is still visible in his forehead – He was carried to a Hospital, where he remained till the spring – When he went to Philadelphia, and there remained till the weather became settled, when he recovered enough to return home, he being unfit for further service in the army, by means of his wound.

The following is his clarification letter written to the Pension Board.

State of Pennsylvania, Lancaster County. Before me the subscriber a Justice of the Peace in and for the said county personally appeared Jacob Heffelbower formerly of Dauphin County but now of Lancaster County who lately prepared an application for a pension to the General Government who being duly sworn according to Law doth depose and say (in continuation or amendment of his former declaration made at Harrisburg on the 21st day of August 1833) That after his time of service which was two months in the militia was expired he wanted to go home, that himself and his comred named Feltman where the persuaded to enlist, that the circumstance of their enlistment was there where two men of Captain Smiths company of Grenadiers drounded and they then wanted two men to fill their place, they then got deponent and said Feltman to try on their uniforms and it fited them well, but they where still not willing to enlist, that General Mercer then promised each of them a tract of land if they would enlist and serve out their time, they then agreed and General Mercer enlisted them and gave each of them eight dollars bounty and promised them *forty shillings per month, that he then served as before stated until the Battle of Princeton, one year and eleven months, and that sad Feltman was killed in the commencement of said battle, the ball having passed though his breast some time before deponent was wounded (as before stated) that each man in his company was numbered on the plate of his cap, deponents number was 25 and Feltman was 10, and the roll was always called by number, that when he was wounded there was five months back pay due him which he never received, that he was so much disabled and lost all his upper teeth from his said wound, that he was glad when he got permission to go home, that after he got back to Lancaster he was one year under the hands of Doctor Adam Kuhn, his father and mother having died while he was in service, that his uncle then fetched him and kept him for three years, deponent laboring for his board and clothing until he got well.*

Author's note – Jacob Hefflebower's pension application was approved and he received $23.33 per year for 7 months service as a private.

There is no mention of his date of passing.

In a statement from his lawyer to the Pension Board, his lawyer states that Jacob's friends find Jacob to be consistent with his stories, and have heard them for many years. They also say that he is somewhat inclined to simplicity and they think this is caused by the bullet wound in his forehead, that is the size of a cent and about a ¼ inch deep.

George Heinish

We head back to the western frontier and the threat of Indian attack on settlers pushing into the interior of Pennsylvania. Mr. Heinish lays out a nice chronological order of his service from 1777 through to 1781. In doing so he describes a very important encounter with Indians in 1781 that really isn't listed anywhere as a major fight, but it had a huge impact on the settlers at the time. There are reports that as many as 30 militiamen and volunteers were killed in the attack, which Mr. Heinish describes. This extremely high casualty rate struck fear into everyone in the area and they thought there was a really big Indian incursion underway. But in reality, it was a relatively small band of Indians who just inflicted a huge number of casualties.

His subsequent account of going after a group of Indians who attacked an outpost documents the brutality and dangers of western expansion.

State of Pennsylvania, Bedford County. On this 28th day of January in the year of our Lord one thousand eight hundred *and thirty five, personally appeared in open court, before the Honorable Alexander Thompson, president and his associate judges of the Court of Common Pleas of said county of Bedford, now sitting George Heinish, a resident of Providence Township in the county of Bedford and state of Pennsylvania aged seventy three years, who being first duly sworn according to law, doth, on his oath, make the following declaration, in order to obtain the benefit of the act of Congress passed June 7th 1832.*

That he entered the service of the United States and served as follows.

That in the year 1777 the Indians becoming troublesome to the inhabitants of Bedford County in as much that they were compelled to leave their houses and farms and take refuge in forts, this applicant with a number of others engaged in military duty at Pipers Fort in said county where he continued in active service standing guard at night and in the day time scouting about through the country. And that he was engaged in such duty three months during said year. That in the year 1778 the Indians again commencing their depredations, the inhabitants as the season before, were compelled again to go to the fort and remain there during the summer and that applicant was also engaged as during the previous summer in doing garrison duty and spying and scouting after the Indians – and during said year served as before three months. That during these two seasons he was not under the command of any officer particularly, but acted under the orders and direction of Captain Piper.

That in the year 1779 applicant was enrolled in the militia of said county under the command of Captain Covalt for two months, which term he served at garrison before mentioned doing constant and active duty.

That in the year 1780 he was enrolled for three months in the militia of said county under the command of Captain Samuel Moore. That during said term they were engaged very frequently through the county and sometimes to a considerable distance. He was three months engaged.

That in the year 1781 Captain Young of Cumberland County was stationed at Frankstown Fort with about sixty men, and about the first of June of that year sent out word to Captain Moore at Pipers Fort to bring as many men as he could procure to Frankstown, and that he (Young) would take as many of his men and they would march north to Clearfield – that on the sixth of the same month applicant having volunteered his services marched from Pipers Fort under the command of Captain Moore for Frankstown with two other men, Hugh Means and Henry Lauthinger, that on the way they were joined by a number of men from Bedford and other places, forming a company of about thirty men – that when they arrived at Frankstown Captain Young informed them that he had no provisions and could not send any of his men with them – When they received this information Captain Moore said to his company that if they were willing they would march themselves to Clearfield and ascertain whether the Indians had been in or were coming in – the men having expressed their willingness to go; They marched from the fort on the morning of the ninth of June, being Sunday, towards the Gap the advanced guard returned to the main body and reported that he had seen eleven fires, and near them a bear which appeared to have been just killed and skinned – The main body then marched up to the fires and halted, and whilst standing there consulting, the Indians, said to be about sixty four in number, rushed out from behind an older thicket in two divisions and attempted to surround them. The company upon appearance of the Indians ran to the trees and when they saw they were likely to be surrounded began to retreat – Applicant recollects that during the battle he saw an Indian sitting down aiming at one of their company, but before the Indian fired, he shot at him, having aimed at his body, and saw him roll over into the bed of a small stream on the bank of which he was sitting – he saw the Indian afterwards trying to crawl out but he did not succeed – the blood at the same time was gushing from his body in a stream apparently as thick as his finger. Applicant saw one of his companions, Joseph Martin shot down by his side, and on looking around saw the Indian jump behind a large white pine. After the battle he returned to Frankstown Fort with Hugh Means, who had his wrist broken, and when they reached the fort, they ascertained that one half of their number including Captain Boyd and Moore were killed and taken prisoner. The survivors next day returned to the battleground where they found seven of their comrades dead, upon whose bodies they threw logs and clumps, not having any implements to bury them with. Applicant afterwards returned home, having been out eight days – Applicant further reflects that during his previous services, on one occasion information was brought to Pipers Fort that a house at which Captain Philips with some men had been stationed in Woodcock Valley, had been attacked by Indians and burned, and that he with seven others started in pursuit. They came to the house, found it burned as stated and *started thence on the trail of the Indians. And after following probably half a mile came to a spot near Tussey's Mountain where they found ten of Phillips men killed, scalped and stripped. Two of the men, brothers, had been tied together to a sapling and shot through with arrows, another had been killed by having been stuck in the neck like a hog – Applicant once also rode from Pipers Fort over Tussey's Mountain after night to Moore's Garrison to give information that the Indians had come in, and went back again the same night.*

Author's note – George Heinish's pension application was rejected without explanation.

There is no mention of his date of passing.

John Hoge

This pension application is very interesting because of his detailing of the parole system. As you read his claim, you'll see he is offered two different versions of parole. After accepting one, he goes home and is still considered a prisoner, only serving his time at home. And when he is released from his parole contract, he goes and collects his back pay for being on parole at home!

It is also interesting because he was granted an initial pension under the 1818 act, but it was taken away after the passing of the 1820 act. This later act, like the initial act, had requirements limiting the amount of assets and income you could have to draw a pension. These were modified again in 1832 which enabled Mr. Hoge to re-apply and draw a pension again.

Pennsylvania Beaver County. On the Third day of September 1832 Personally appeared in open court before the Court of Common Pleas now sitting afore said county, John Hoge, a resident of said county of Beaver aged Eighty-Two yeas who being duly sworn according to law doth on his oath make the *following declaration in order to obtain the benefit of the act of Congress passed June 7th 1832.*

That he was commissioned a lieutenant in the regular army on the 6th day of February 1776. The commission was signed by John Hancock, President of Congress. He was placed under Colonel William Irwin of the 6th Pennsylvania Regiment. He was furnished with money to recruit by the colonel and was engaged in that service in York County Pennsylvania before the regiment was filled, when he was placed under Captain Moses McClean and marched from York County to Philadelphia in May 1776 and from thence to New York and thence by water to Albany and from thence to the Half Moon twelve miles above Albany, where he was placed as commander of forty eight picket men with two sergeants and two corporals, called a Flanking Light Infantry Company. From that place went by water across Lake George, from thence down Lake Champlain to Fort St. John, from thence to Fort Chombalee,

Author's note – Fort Chambly

from thence to the mouth of Sorell on the St. Lawrence and was there under the command of General Sullivan. At that place they built a log fort. From that place the regiment to which this deponent belonged, the second and forth regiments; the second commanded by Colonel St. Clair, the fourth by Colonel Wayne, went down the St. Lawrence forty-five miles to attack the British at the Three Rivers. When the army arrived at the Three Rivers they attacked the enemy but were repulsed and forced to retreat. They then returned to General Sullivans camp at the mouth of Sorell. From that place the whole army were marched and taken by water to an island in Lake Champlain called the Asle Aux Noix, there they were attacked by Canadians and Indians. When deponent and a party were out on scout and he was taken prisoner after an engagement in which Captain Adams and Ensign Cubberren and two privates were killed, at that skirmish Captain McClean was wounded.

Deponent was sent as a prisoner under guard to Fort St. Johns where the British had taken possession, and thence to Chombalee and thence to Montreal and was there handed over to General Carleton by the Indians, with his captain (McClean) who was also a prisoner with several others. Who were all bound with cords, and during which time they considered their lives from the threats and menaces of the Indians in imminent danger, but when delivered over to the British they considered themselves more secure.

From that place the prisoners were all taken to Quebec, deponent was kept there about two months when a Parole was presented for them to sign with a stipulation that they would not take up arms again during the war. This the Virginians and Pennsylvanians refused to do. On two days afterwards they were presented with another Parole with a condition that they might go home and that if called upon to do service before an exchange of prisoners, that then they must surrender themselves again as prisoners. This Parole deponent with the other prisoners signed. They were then sent to New York and arrived there about the last of September. From thence they marched to Philadelphia and drew three months pay. Deponent was there made Paymaster. From there deponent went home and stayed until May 1779 when he was informed by Captain McClean, who was his captain, that he was exchanged. He *lived then in Cumberland County near Carlisle Pennsylvania. On receiving this information, he went immediately to Head Quarters close to Sommerset Court House in the state of jersey and reported himself ready for duty and called upon the colonel of the 6th Regiment, which was then commanded by Colonel Hays, and demanded his rank. He was informed that the regiment was full, but that the first vacancy he would be called on. He returned home and was never again sent for nor had no further notice on the subject.*

At the time he went to Head Quarters and reported himself after his exchange, he drew his pay up to the time he was exchanged. General Washington signed the warrant for his pay, which he drew in Continental paper, which he kept until it died in his hand.

Author's note – Continental paper money became worthless after the war.

After the close of the war he took his commission to the controlers office at Philadelphia to draw a years pay which was due him, he left his commission there and never again seen it.

Deponent relinquishes every claim whatever to a pension or annuity except the present and declares that his name is not on the pension roll of any agency of any state. That under the Act of Congress of 1818 he was placed on the Pension List, but by the Act of 1820 his name was stricken off on the grounds that he had to much property to intitle him to the pension of that act.

Author's note – John Hoge's pension application was approved and he received $320 per year for service as a lieutenant.

He passed away on the 10th of May 1834.

James Huston

In his extremely detailed pension application Mr. Huston does an amazing job recalling names and events that I think you'll find very impressive for a seventy-six year old. His application reads more like a short story and is a rare gem in what it was like to be fighting on the frontier of Pennsylvania.

State of Pennsylvania County of Indiana. On this twenty fifth day of March AD 1834 personally appeared in open court before the Judges of the Court of Common Pleas in and for Indiana County in the state of Pennsylvania now sitting James Huston a resident of Centre Township Indiana aged now seventy six years – who being first duly sworn according to law doth on his oath make the following declaration in order to obtain the benefit of the act of Congress passed June 7th AD 1832.

That he entered the service of the United States under the following named officers and served as herein stated – that in the beginning of the year 1777 he resided in that part of Westmoreland County Pennsylvania which is now Indiana

County and that about the fifth day of April in the year 1777 he volunteered in a company of Rangers commanded by Captain John Pumroy, he First Lieutenant was John Hopkins, Second Lieutenant William Lemon, Ensign Joseph Hopkins – The company rendezvoused at James Ramsey's who resided at that home about two miles north east of where the borough of Indiana is now located, that early in the month of June 1777 Captain John Pumroy was elected Colonel – And the command of the company was handed to John Hopkins, William Lemon was First Lieutenant, John Lemon was Second Lieutenant and Joseph Hopkins continued Ensign – That on the fifth or sixth of August 1777 Deponent was discharged having served four months in this tour – That during the whole four months the company was stationed at James Ramsey's whose house was converted into a block house and Deponent was almost constantly engaged in scouting parties against the Indians for the protection of the frontier, being seldom more than a day at one time in the station and during the time the major part of the company were twice as far as Kittanning, once under Captain Pumroy and once during the command of Captain Hopkins – but Deponent was not engaged during these four months in any skirmish with the Indians, although they frequently saw what they supposed to be their trails – James Ramsey where the company was at, furnished the company with rations, during the whole time Deponent did not ask for nor received any written discharge.

That about a week after Deponent was discharged from Hopkins command on or about the thirteenth day of August 1777 Deponent volunteered in a company of militia commanded by Captain Samuel Dickson, Robert Rayburn was

First Lieutenant, John Miller Second Lieutenant and Robert Mitchell Ensign. The company was formed from the militia residing within the borough of Captain Dickson's company volunteering as Rangers and put under pay and rations under a requisition from Archibald Souyhey, County Lieutenant – The company was stationed at Samuel Dickson's about a mile from Campbell's Mile on Black Lick Creek, then Westmoreland County, now Indiana County – Colonel Campbell was commissary and being a source of salt, Lieutenant Rayburn and Deponent were sent for some salt in kegs which Deponent had left at his place covered with flax in an outhouse – While Deponent was gone for the salt – Colonel (afterwards general) Campbell, David Dickson (a brother of Captain Samuel Dickson), John Gibson and Rannel Laughlin went to the plantation of Rannel Laughlin to ascertain whether that would be a suitable place for a station, being six or seven miles further out than Dickson's – As Deponent was on his return to the station with the salt he killed a deer and took the meat past the station where he left the salt with his mother and sister at Wallace's Fort and stayed the night at the fort – He returned to the station next morning returning there that day – And Colonel Campbell and those who were with him not returning according to their purpose – The next morning Captain Dickson, Levi Gibson, and this Deponent started out to Laughlin's plantation to learn what had become of them – When Gibson got near the house and opposite the door he turned round startled and ran back and said "that was enough" Captain Dickson called on him several times to stop but he continue on and Captain Dickson and the Deponent pursued him and when they came up with him he said he had seen Campbell's dog dead at the door of the house – Captain

Dickson proposed to go back to the house but Levi Gibson refused and continued on his way to the station and Dickson and Deponent went with him and the company evacuated the station and went to Wallace's Fort in Westmoreland County and expresses were sent out for men from other forts and then was a group of one hundred men collected and they started to discern what had become of Colonel Campbell and his companions – Captain Dickson commanded the advance guard which consisted of Deponent and eleven men – When they got near to Dickson's Station they heard a dog bark – And the captain sent Lieutenant Rayburn and Deponent round through the field to come up behind the house and sent word to the main body to come up, intending to surround the house – The main body came on so fast that the Indians got the alarm and before Rayburn and Deponent had got to the place assigned them by Captain Dickson (having a considerable circuit to make) Joseph Campbell and John Scott, two of the main body came between them and the Indians and got before them and fired at the Indians who were in a piece of ground – Dickson had cleared a meadow, Joseph Campbell was reloading his rifle when he was shot by an Indian – and ran about fifty yards and fell dead – The Indians then started to run and Colonel Ramsey shot at one through a fence, which brought him to his hands and knees, but he rose again and ran off – The troops pursued them about three quarters of a mile but the Indians escaped and the troops returned to Dickson Station and buried Joseph Campbell and continued there all night and next morning went to Rannel Laughlin's house which the troops surrounded but it was empty and a letter was found sticking in the door which had been written by some of the British to the inhabitants urging them to give up and return to their allegiance – Colonel Campbell had written on the letter that they were all taken prisoners but were well – The greater part of the men then went home – but Major Wilkins, Captain Dickson and the Deponent with a scouting party went across the county to two Lick Creek to endeavor to make further discoveries – but not making any they returned to Wallace's Fort – Dickson's Station being then abandoned – And Deponent continued from that time to be engaged on scouting parties through the county until Wallace's Fort was attacked by the Indians which was in the month of October 1777 about the middle or later end of the month – In the morning of that day Major Wilkins and William Campbell (a brother of the colonel) had left the barn where they had been feeding their horses, the Indians came in and shot a calf and a black sheep – Deponent not knowing that Wilkens and Campbell had returned to the fort, ran out of the fort towards the barn when Captain Dickson called to know where he was going – Deponent replied that he was going to the barn that he expected the Indians had shot Campbell and Wilkins – Captain Dickson told him to go back that he had seen them go into the fort – Deponent and Dickson then both went into the fort where he found Campbell and Wilkins whom he had not observed before, they being in Chapman cabin – Major Wilkins then stationed every man at his port hole round the sides of the fort – only a few men were stationed as the fourth side – that being covered by Wallace's Mill in which there were ten men stationed and Captain Dickson went out and took the command of them – the mill was grinding corn at that time for the troops there being no whet – about an hour after this nine Indians came out of a thicket which was between the fort and the barn and coming near the fort raised their guns and fire toward it and raised the whoop and run off beyond the barn, in about half an hour fifteen Indians came out of the same thicket and repeated the same evolution – the men in the ort wanted to go out and fight them – but Major Wilkens refused permission as he said the Indians only wanted to get them out by that maneuver - Shorty after a white man came up the trail of the mill with a red flag hoisted in a pole, he got bedside a buttonwood tree and John Donehy fired from the fort and struck him in the right side and the ball went through him – he shifted round the tree – when he was struck in the breast from a shot from the mill and fell dead – Deponent afterwards understood from Captain Dickson that it was he that shot from the mill – There was another man with the white man who made his escape when the man was shot but they couldn't discover whether he was a white man or an Indian.

Immediately after this man made his escape the Indians commenced a general fire upon the fort which was returned. In the fort they could hear the bullets of the Indians strike against the stockade. But they were so far off that very few of the bullets penetrated into the wood – After the firing was over the Indians took five horses out of the barn – Three of which belonged to Major Wilkins, one to William Campbell and a stud horse belonging to Richard Wallace – After the Indians had gone Major Wilkins called down to the mill to know whether they wanted any men from the fort to assist in taking in the white man who had been shot – they replied from the mill that they did and Deponent and some others ran down from the fort to the mill and those in the mill refused to go out until Deponent and Edward Cahill ran beyond where the man lay to the top of a hill to see if the Indians were gone – when those from the mill hauled in the body – By a subsequent account which was received it was stated that the Indians were three hundred strong when they attacked the fort – Deponent from this time continued to be engaged in scouting parties continually so long as Captain Dickson continued in command while the Indians were burning the houses and laying waste the plantations, but he never came up with them but once when a scout was out commanded by Major Wilkins and they were accompanied by a scout commanded by Captains John Hinkston from Fort Palmer, Ligonier Valley, Westmoreland County Pennsylvania, while on scout they followed an Indian trail across Cowanshannock Creek in what is now Armstrong County Pennsylvania, Captain Hinkston – Daniel McClinlock and Deponent were some in advance when they had got about three miles beyond Cowanshannock they spied the Indians kindling a fire and they fell back and informed the main body of it and they came to the top of the hill and then returned to a deep hollow where they kindled their fires and staid there late, they allowed the Indians would be in their first sleep – The company were then divided into two detachments so as to surround the Indians and tomahawk them without giving any alarm – But one of Hinkston's men named Wilson when they got near the Indians shot one of them dead as he lay – when the rest started up and endeavored to make their escape – three more of them were killed and a fifth one made his escape – About the middle of February 1778 Captain Dickson left this company and Captain Andrew Lower assumed the command and some additional men joined and Lieutenant John Miller died about this time and the other officers remained as under Dickson – While under command of Captain Lower the company was engaged in guarding the people who had grain to thrash and continued engaged in this service two months – about the middle of April Deponent was discharged as well much of the company – Deponent having been engaged eight months in the tour under Captain Dickson and Lower during the whole of which time he held the appointment of first sergeant – but he never asked for nor received a written discharge nor does he believe it was customary to give any.

The settlement in Westmoreland County being broken up Deponent took his mother down to Conococheague in what is now Franklin County Pennsylvania where he was drafted in the later end of August 1781 in a company of militia commanded by Captain John McConnell, David Shields was lieutenant and the ensigns name Deponent cannot recollect – Colonel Thomas Johnson and Major William McFarland of the Big Spring were with the company and had the command when they marched to Standing Stone in Juniata River, now Huntingdon County Pennsylvania, and the company lay there about two weeks going out occasionally to assist the county people in pulling in their grain and hay – After which the company marched to Frankstown settlement in what is now Huntingdon County and lay at Peter Litus's and then met two other companies of militia, one of which was commanded by Captain Samuel Holliday Does not recollect who commanded the other company – The militia lay there for four weeks watching the gap of the Allegheny Mountains where the Bedford scouts were killed by the Indians – The provisions ran low and Captain McConnell and Captain Holiday's company were marched back to the Standing Stone – The other company was sent to Rickett's Fort while the militia lay at the Standing Stone Captain McConnell was directed to detail a number of men from both companies and set out upon a scout – Deponent was one of the men selected and they left the Standing Stone, went up through Frankstown settlement, up what is now Calter Blue Gap and crossed the Allegheny Mountains to the head waters of Two Lick Creek and returned back through the gap where the Bedford scout had been killed to the Standing Stone and shortly afterwards the company was discharged, which was in the later end of October. Deponent having been engaged on this tour two months. Deponent did not ask for nor received any written discharge.

Author's note – James Huston's pension application was approved and he received $40.66 per year for service as a private.

He passed away on the 16th of September 1841.

William Jenkinson

We head back to the fighting in the southern campaign as William Jenkinson lays out a nice story of his service which covers two and a half years. He becomes an artillery man for part of his service and I think it adds to the uniqueness of his claim.

There seems to be some confusion with his surname, which he states in a clarification letter declaring that his name is, and has always been, Jenkinson. But the orderly sergeant, who he names as Moorebeggs, always called roll referring to him as Jenkins. He even goes as far as to make his pension claim in the name of Jenkins and signs the claim Jenkins. In the end he is granted a pension as Jenkinson who served under the name Jenkins.

State of Pennsylvania, Westmoreland County. On this 13th day of January 1834 personally appeared in open court before the judges of the Court of Common Pleas of Westmoreland County now sitting William Jenkins of Unity Township in said county aged 79 years who being duly sworn according to law, doth on his oath make the following declaration in order to obtain the benefit of the act of Congress passed 7th June 1832.

That he entered the service of the United States in the spring of 1776 in a company of Rangers commanded by Captain McCoy of Cumberland County Pennsylvania. We marched from Old Town to Potter's Fort in Penn's Valley, we were employed here in ranging the country between this place and Clark's Fort in Buffalo Valley, keeping off the hostile Indians and were under the command of Colonel Welkner. I received a discharge from our captain, but cannot now tell what has become of it – In this year I served a tour of seven months – In the spring of 1777 I served a tour of two months duty as a substitute for Hugh McClelland in a company of militia commanded by Captain Wilson at a place called McCormick's Blockhouse in Standing Stone Valley in Bedford County – In the month of June 1781 I enlisted in the 4th Pennsylvania Regiment of Continental Troops then stationed at Carlisle, Cumberland County, commanded by Colonel Graig. I served in Captain Thomas Campbell's company. We marched from Carlisle to Little York where we remained for some time drilling the regiment. We were marched from Little York to 4 Mile Run near Philadelphia and from thence to Philadelphia where we embarked and sailed to Christeen Bridge where we were landed and marched to the Head of Elk in the state of Maryland. We there went aboard of several small vessels and sailed to Baltimore from which place the company to which I belonged and another company embarked aboard a small galley and sailed under convoy of the Washington Privateer to the mouth of James River in Virginia. We passed through the French fleet which were lying off York harbor blockading

Cornwallis's army – We landed and joined the American's before Yorktown. That day or next a draft was made on our Regiment to fill up an artillery Company, two men were drafted from each company, I was drafted from Captain Campbell's company and was enrolled in Captain Ferguson's company of artillery. The regiment was commanded by Colonel Proctor. Our lieutenant colonel was Tarleton. On the 9th or 10th of October I was stationed on a battery on one side of which the guns were directed against a battery that was erected between us and Yorktown and a mortar and two howitzers on another side were directed against the British vessels lying in the river. We burned two of their ships in the river. I was employed both days at one of the howitzers – A few days afterwards Cornwallis surrendered. We crossed James River and lay there for some time. The Pennsylvania Troops under the command of General Wayne and General St. Clair marched to South Carolina where we joined General Greene's army on Christmas day 1781 at a place called Ponpon. The advance guard of General Greene's army was at Bacon's Bridge. We here encamped on the Cypress Hills and on James Island. A bridge of boats was thrown across Cypress River. We crossed and destroyed a fort belonging to the British and brought off some of their cannon and returned to camp on Cypress Hills. The British about this time sent a number of Row galleys up the Gumber River. A detachment was immediately sent from our camp commanded by General Guess of Maryland. The night got so dark we could not proceed and at early dawn we heard the militia that had got ahead of us firing. General Guess ordered Captain Ferguson with his cannons of artillery and a detachment of the Delaware Troops, the whole were under the command of Colonel Lawrence. We marched on and had to *cross a pole bridge and reached a battery on the banks of the river where we planted our guns. One galley had passed down and another was within range. We poured a volley of canister shot into her. They immediately took to their oars and pulled up the river. At this place the river made a bend and the galley that had passed down cast anchor and began to play on us very briskly. We played at her with a double fortified 12 pounder. At this time Colonel Lawrence came galloping up and commanded Captain Ferguson to retreat immediately for that the British had effected a landing up the river, and that General Guess could not get forward to our support. We immediately limbered up our pieces and retreated and the militia ran off entirely. We were met by Major Bell, aide camp to General Guess who told Colonel Lawrence to commence the attack on the British. We immediately unlimbered our pieces. I was stationed at a howitzer. We loaded with canister shot. The British came up under cover of two bayous that ran out of the swamp into the river and were covered by low brushwood. They fired on us and Colonel Lawrence and his horse were both shot dead at the first fire. We then retreated across the pole bridge. My messmate was taken prisoner and I would have shared the same fate but I got amongst the Delaware Troops who fought like as many devils, but they were terribly cut up. When we got over the pole bridge we met General Guess with his detachment. We then went to camp. We had an engagement after this with the British on the main road to Charleston. They retreated but we did not pursue them far. About this time Lieutenant Colonel Tarleton had brought up clothing for us but in crossing the rivers the clothing had gotten wet and was entirely rotten. And we were nearly naked and had no money and our troops were like to mutiny. A sergeant by the name of George Gosner was shot and so the matter ended. General Wayne or Greene then issued orders for all those who had served out their 18 months to be marched home – We then marched home under the command of Captain Studsburry to Lancaster Pennsylvania and lay there one month before we were discharged. We were then discharged by Colonel Hampton. It was just twenty two months from the time I enlisted at Carlisle till I was discharged at Lancaster. My discharge was in print and mentioned the time I had served over the 18 months and that I had received no pay for any of the time, and I never received one cent for said service. I lost my discharge in Carlisle the year following. I advertised for it but never could hear anything of it.*

Author's note – William Jenkinson's pension application was approved and he received $30.00 per year for 30 months service as a private.

There was no mention of his date of death in is record.

James Johnston

In this lengthy and exceptionally well documented account of his service, James Johnston lays out a vivid account of the life of a soldier who mingled with the likes of George Washington and La Fayette. He is an exceptionally well spoken and educated man who really draws you into his story from the onset. I like to try and give an introduction to these men, but this man clearly lays out a story that speaks for itself. This is a real gem and I think you will agree it is a unique piece of history.

State of Pennsylvania, Allegheny County. On the thirty first day of July in the year of our Lord one thousand eight hundred and thirty two, personally appearing before the Honorable, the Court of Common Pleas of Allegheny County now sitting James Johnston a resident of said county of Allegheny state of Pennsylvania, aged seventy four years, who being first duly sworn according to law, doth on his oath make the following declaration in order to obtain the benefit of the act of Congress passed 7th June A.D. 1832.

That he was born in Chester County, state of Pennsylvania on the 22nd June 1758. He is possessed of no record of this fact nor does he believe that any such record exists. Declarant was living in said county on the farm of his father at the breaking out of the war of the American Revolution – Two of his elder brothers obtained commissions in the army – one Francis Johnston as a colonel and the other Alexander Johnston as a captain in the Pennsylvania Line – Declarant was at no time in the regular service having declined a lieutenancy (obtained for him by his brother Francis as said Francis informed declarant) from unwillingness to leave permanently his aged father of whom he was the youngest son, and the only one remaining at home:

On the 8th June 1776 declarant marched from Chester County to New York with a company of militia to which he had attached himself, commanded by his brother – in – law Doctor John McDowell and forming part of Colonel William Montgomery's Regiment. He was is not aware that there was any formal draft but believes that they proceeded on the emergency of the case; there being in the ranks some of the most respected citizens of the county, amongst whom he remembered Stephen Cochran and Montgomery Kennedy, brother to the surgeon general of the army. The enemy then occupied Staten Island – declarant was stationed on the opposite Jersey shore at a place called the Blazing Stars where was collected a force of about one thousand men composed of Colonel Montgomery's regiment and of another regiment also militia from Lancaster County commanded as declarant thinks by Colonel John Boyd – Declarant recollects one incident which occurred whilst in this position watching the enemy. An

alarm having been given during the night of a movement on the water, an officer was directed to proceed in pursuit and examination. The selected for this purpose, the members of the mess 10 which declarant belonged, and manning a boat with them proceeded (being seven in all including the officer) to execute the order. After a long and ineffectual pursuit, it was found as morning dawned, that the bot had approached close to Staten Island and she was hailed by a guard of the enemy consisting of six men under command of a sergeant and a fire of musguetry opened on her – Declarant and a young gentleman named Thomas Evans were then at the oars – The officer and the four others immediately crouched down to the bottom of the boat – whilst declarant was remonstrating against this conduct as leading inevitably to capture, the boat touched a projecting sand bank and the officer made a spring to the shore, but fell into very deep water, he rose, succeeded in getting ashore, and immediately in the most dastard manner, took to flight, Declarant seized a musguet and taking deliberate aim at the sergeant wounded him severely as was inferred from his immediately dropping his gun and disappearing in the tall grass and weeds, Evens also took aim at and wounded one of the enemy. Taking advantage of the pause this occasioned, the boat was got out and returned to the American camp.
Declarant on this occasion received a slight concussion from a musguet ball. He continued with Captain McDowell's company until September or October 1776 when it returned to Chester County and was discharged from service.

In November 1776 declarant again marched from Chester County as a member of Captain McDowell's company, the regiment being commanded y Colonel Even Evens. They remained in Philadelphia until December 1776 when they were taken up the Delaware in the Row Gallies and landed in the night on the Jersey shore about three miles above the town of Burlington – having in company a body of troops believed to have been regulars from New England. Soon after reaching the point mentioned Generals Putnam, Cadwalader, and Mifflin arrived. Declarant was familiar with that part of the country having formally gone to school in the neighborhood and on hearing inquiries made by these officers as to the route to Bordentown, he stated his intimate knowledge of it and his willingness to take the responsibility of guiding the column – no movement however was made, and the declarant well remembers the chagrin expressed by General Mifflin on ascertaining afterwards that had the troops marched at once in the direction indicated they would have fallen in with a body of Hessians proceeding from Mount Holly to Princeton. On the next morning the whole force proceeded to Bordentown and marched thence the same day to Crosswicks. Here they lay until New Year's Day, 1st January 1777 when an express arrived from General Washington commanding a junction with him at Trenton that night – On beating to arms the New England troops whose term had expired a few weeks before, and who were almost destitute of clothing, declared that they would perform no further service - General Mifflin addressed them in a very animated strain, and finally asserted that all who were willing to march should step forward and give three cheers – Every man of them did so – Declarant particularly remembers the night of General Putnam at this result. After marching all night, Trenton was reached about day light and the force to which declarant belonged was ordered to occupy the Green House of General Dickinson, about one mile above that place, but whilst preparing breakfast the alarm was given of the approach of the enemy. Hastening back to Trenton across the bridge over the Mill they joined the army which was formed into three lines, front, center, and rear. About one o'clock the enemy approached in sight. General Washington who was well supplied with artillery ordered twelve pieces to be placed behind the mill – About three o'clock in the afternoon the enemy came down the streets in solid column when the artillery opened upon them, a most destructive fire, which caused them to recoil – At night preparations were made by General Washington for a retreat, declarant was detailed for the main guard, and shortly after relief he found the army about to move. Many of the soldiers thinking they were about to be led against the enemy threw away their knapsacks. General Washington enquired at the rear of the column who commanded here? Major Bell of Colonel Evens regiment announced himself as in command and received orders to remain behind for two hours, carefully observing the enemy – Declarant was with Major Bell and was sent forward by him with a party to reconnoiter the enemy moving carefully forward. They had a full view of the Hessians sitting round their fires smoking their pipes. A sentinel challenged declarant and the others dropped to the ground, and lay quiet until the sentinel was heard t resume his walking. They then cautiously retrograded and made report. At the expiration of the two hours Major Bell proceeded to join the main army. The advance had an engagement with the enemy at Princeton, but the force under Major Bell was not in time to take part in it.

In May 1777 a corps consisting of drafts from various parts of Chester County and amounting in the whole to about six hundred men marched under the command of Colonel John Hannem. Declarant was specially appointed by Colonel Hannem adjutant of this corps. He believes that he held a written appointment from Colonel Hannem, but it is no longer in existence. The corps proceeded to the town of Chester and subsequently to Billingsport where it remained in service until sometime in August 1777. During the whole of this period declarant was in the active and auspicious discharge of the duties of his appointment in every matter connected with the drill and other military duties, and may remark that the corps was known and distinguished at the time for its state of discipline and efficiency. Major Hartman, a German, who had been in foreign service, took particular pains and a very friendly interest in instructing declarant in the minutest points of duty, Whilst the corps lay at Chester General Potter, who commanded a body of militia from Weston Pennsylvania urged declarant to act as his aid and on his consent imposed on him at once the duty of arresting an officer who commanded a corps composed principally of Germans, from Northampton County who had committed some gross military offense. Declarant proceeded immediately to demand a surrender of the offender's sword. He consented, but a large body of his men instead unanimously on its restoration. Declarant drew his sword and was about to meet the urgency in a fitting manner, when a party of his own corps rushed to the spot and could with difficulty restrain them from using their arms on the offenders. Whilst the corps lay at Chester the declarant was ordered to convene all the officers at 12'oclock noon at Curling's Tavern. On the arrival of the hour it was found that the object was to meet three commissioners sent from Philadelphia for the purpose of administering an oath of *allegiance to the new government. Colonel Hannem remarked to the commissioners that the subject was of great importance and delicacy, and that it would require much time for deliberation; the same was said by others and was becoming the sentiment of the room. The declarant came forward and earnestly insisted on the importance of enabling the government at once, to know their true and staunchest friends; the deponent declared his readiness to take the oath, which was administered to him accordingly – and before the commissioners left the house it was taken by every officer.*

Shortly after declarant's return from this tour of duty the enemy in great force ascended the Chesapeake – They landed at Turkey Point on a Saturday. On Sunday morning at 11'oclock a hasty letter dated Elkton reached declarant from Colonel Henry Hollingsworth, deputy quartermaster general, stating his fear that the public stores would fall into the hands of the enemy – and entreated declarant to use every possible exertion to aid in effecting their immediate removal. Before 3'oclock in the afternoon between 30 & 40 wagons engaged by declarant were on their way for that purpose, and before night fall declarant was himself at Elkton in conference with Colonel Hollingsworth. The colonel spoke of the house and out houses of declarant's father as fit places of deposit. Declarant feared that they would be in the route of the enemy and moreover that his father's well served and enthusiastic feelings as a Whig and the freedom with which he delivered his sentiments on all occasions would be the very means of attracting notice and resentment. It was finally decided however to send the stores therein five hundred barrels of flour, 100 barrels of rum, a great quantity of pork, salt, molasses, and other articles were accordingly dispatched. By Tuesday at 10'oclock everything was carried off; and not until then did declarant close his eyes, subsequently to the receipt of Colonel Hollingsworth's letter – On Thursday the British army marched through Elkton – Declarant repaired to the head quarters of General Washington, where were his two brothers, Francis and Alexander. Being well mounted, he with some other young men reconnoitered closely the enemy's position at Grey's Hill – The enemy were not provided with cavalry, but a party of mounted men closely pursued declarant whilst thus engaged. Declarant had occasion to return to his father's house – but the day before the Battle of Brandywine, he rejoined the army. He was mounted and took a position with the force under General Potter – owing to the unexpected turn which the battle took General Potter, who was stationed below the lower ford, had no active part in the conflict of the day – The public stores already referred to, remained in safety at the house of declarant's father, and at the express insistence and request of Colonel Hollingsworth declarant issued them during the fall and winter of 1777 and spring of 1778 to the American army at Valley Forge – Great caution was necessary on account of the occupation of Philadelphia by the enemy – The course pursued was by Cochran's Tavern to the Lancaster Road. Declarant had great difficulty with the wagoner's. He also supplied the troops of General Gest from Maryland, who came on a few days before the surprise of General Wayne at the Paoli and greater part of whom dispersed immediately afterwards. The whole was done under the orders of Colonel Hollingsworth. Declarant thinks he had a warrant in writing from that officer – He kept a regular account of the issues and made return thereof to Colonel H.

In the month of March 1781 troops were passing to the southward to act under the Marques de La Fayette in repelling the incursion of the British under Arnold and Phillips – Declarant joined them near Wilmington in the state of Delaware and whilst crossing the Susquehanna was introduced to the Marques to whom he expressed a wish, being well mounted, to be attached to any cavalry corps, that might be employed – The Marques yielded a ready and cheerful assent. At Fredericksburg declarant obtained permission from the Marquis to turn aside for one day to visit a brother residing about 25 miles below. Whilst returning next morning declarant met a party of Americans who had fled from Richmond and who were engaged in sawing apart some beautiful pieces of cannon presented to the United States by the French government, and bearing the arms of France. This was with a view to render them useless in the event of their falling into the enemy's hands. Declarant remonstrated and advised that they should be buried, but was answered that the orders were explicit.. On reaching Richmond he reported the scene he had witnessed to the Marques La Fayette and General Nelson. The former showed considerable emotion and the latter was exceedingly indignant. – Adverting to the subject of declarant's wish for employment, the Marques said "There are a hundred young gentleman Virginians desiring of serving as officers, but no one is willing to go into the ranks." Declarant expressed the cheerfulness with which he would occupy any position even under an officer of the regulars. One incident of that day is strongly impressed on declarant's memory. He was invited by the Marques to dinner at a small log house opposite the ferry below the Fall's. Declarant happened to be so laced at the table as to be able to see what was passing out of doors, and he called attention to the fact that General Phillip's and Arnold had advanced along the beach and were making an examination with spy glasses whilst their servants held their horses. Immediately after there was a bustle at the door occasioned by five riflemen in hunting shirts and moccasins who eagerly solicited permission to shoot down to a point from which they felt sure they could pick off these officers. The Marquis refused his sanction declaring that he would meet the enemy openly in the field, but would authorize nothing like assassination. This refusal excited great dissatisfaction, which was expressed amongst the rest, by his aid Major McPherson. Declarant was attached by the Marques to the troops of Captain Reed of Virginia. The enemy retired down the river to Petersburgh. Declarant served with Captain Reed until a call of business and the absence of any immediate prospect of active duty, induced his return to Pennsylvania, which he reached after an absence of two months.

Author's note – James Johnston's pension application was approved and he received $70.00 per year for 3 months service as an adjutant and 21 months service as a private.

There was no mention of his date of death in is record.

James Josiah

Although James Josiah's story is short, it documents a major milestone in American naval history. He served as first lieutenant aboard the brig Andrew Doria, which was presented with the first salute of and American naval vessel from a foreign country. Although he just missed this moment in history, due to being taken prisoner, he does participate in much of the Andrew Doria's success leading up to the salute.

The historical salute was given by the Caribbean, or West Indies, country of St. Eustatius on the 16th of November 1776. As Andrew Doria entered the foreign port, she fired her guns in salute, and received the same from the fort protecting the harbor.[9]

James Josiah a citizen of the United States, resident in the County of Philadelphia, State of Pennsylvania, aged sixty eight years, on his oath declare: That in or about the month of October One Thousand Seven Hundred and Seventy Five he was appointed first lieutenant of the Brig Andrew Doria fitted out at Philadelphia and commanded by Captain Nicholas

Biddle. That he sailed with the fleet under Commodore Hopkins on the expedition against New Providence and after the surrender of that place returned in the said Brig with the fleet to New London.

Author's note - New Providence is an island in the Bahamas.

That he sailed in the same vessel as first lieutenant on a cruise upon the banks of Newfoundland, and the Andrew Doria having captured two transport ships with Scotch troops on board, the command of one of them was given to the deponent with orders to proceed to Newport or New London, But off Nantucket he was taken by the British Frigate Cerberus and detained in her three months, was afterwards put on board the prison ship Whitby at New York, where he remained near five months when he was exchanged on the 24th December 1776.

Author's note - The Andrew Doria ran into the British transport vessels Oxford and Crawford loaded with Scottish troops. James Josiah was placed aboard Crawford.[9]

The ill treatment received by this deponent while in the hands of the enemy was made the subject of complaint by Captain Biddle as well to the British admiral at New York and to the American government at the council of congress of 7 August 1776, there the notice taken of the subject by that body. On the 20th August 1776 while a prisoner the deponent was promoted to the rank of captain in the navy of the United States and appointed to the command of the ship Champion lying in Philadelphia, and was engaged in her at the siege of Fort Mifflin and Red Bank.

Author's note - Champion was armed with 8 guns.

After the evacuation of the fort, the said ship with others was burnt by order of a council of war and the deponent proceeded with his crew by boat to Bordentown where he fell out of the service and entered the private service.

Author's note - As the British closed in to capture Philadelphia, those vessels that were unable to break out were burnt to prevent capture.

After the evacuation of Philadelphia by the British in June 1778, the deponent obtained a furlough to command a privateer and made several voyages in the private service, having orders from the navy board to report himself on his return to port, which he conformed to do until the Peace, never having been discharged from the service. When the deponent entered the private service he gave up his commission applicable to the general order of the navy board and did not again receive it having continued in said service until the conclusion of the war, though able and ready to do duty when called upon by government.

Author's note – After the burning of their vessels several of the captains were given furloughs to enter private service as long as they checked in with the navy whenever they came into port to see if they were needed for active duty service.

James Josiah's pension application was approved and he received $20.00 per month for service as a captain.

He passed away on the 18th of September 1820.

His pension claim deposition was given on the 14th of January 1820, so he only was able to draw about 8 months of that pension. But his wife Elizabeth, who was 79 years old at the time, filed a claim for widows later on in 1843 and began to draw $552.00 per year on his pension.

Frederick Kalehoft

This story is a great example of the indentured servants who fought in the war. I came across these occasionally, although indentured servants were common at the time. People like Frederick came to America on voyages paid by people - Frederick refers to as his master - who lived in the colonies. And as he states in his claim, he is in service to his master for 4 years to pay off his passage. During his indentured service he substitutes for his master and his master's son, who were both drafted into the militia. But I have read pension claims that state that indentured people substituted for friends of their masters. More than likely to pay off a debt owed by their master.

State of Pennsylvania, Columbia County. On the eighth day of November A.D. 1832 personally appeared in open Court of Common Pleas before the Honorable Seth Chapman esquire, President and his associate Judges of the same court now sitting Frederick Kalehoft or Colehoof, a resident of Catawissa Township in Columbia County aforesaid state aged seventy two *years who being duly sworn according to the low, doth on his oath make the following declaration in order to obtain the benefits of the act of Congress passed June 7th 1832.*

That he entered the service of the United States under the following named officers and served as herein stated – That in the year 1774 he arrived at Philadelphia from Germany, then thirteen years of age – was bound for four years to Christopher Garret to pay for his passage – Christopher Garret, his master resided at Pottsgrove – In August 1776 deponent's master was drafted into a militia company at Pottsgrove, Pennsylvania. Said company was commanded by Peter Richards captain – deponent entered said company as a substitute for his master, Christopher Garret – was marched to Amboy in New Jersey, served in that company two weeks when application was made by the government officers for volunteers, a draft of the militia men then there in the service was made – James Garret, a son od deponents master was then in the company deponent was, and was drafted into the flying camp and at his request deponent joined the flying camp as his substitute and was attached to a company commanded by Captain Joseph Huster, former governor of Pennsylvania.

Author's note – Joseph Hiester was the governor of Pennsylvania from 1820 to 1824.

Was then marched to Long Island, was in the Battle of Long Island, our Captain Huster and our Lieutenant Christopher Braldy were taken prisoners, General Sterling was in the battle. After the battle deponent joined General Potter's Brigade – Captain Hurlay of Lancaster was deponent's captain – deponent was then marched to New York, crossed there into

New Jersey, from there through the Jerseys, was at the Battle of Trenton when the Hessians were taken, Washington was there – General Potter was then also in command of our brigade. After the Battle of Trenton General Potter crossed the Delaware and we were marched down the Delaware to Wilmington, the British were coming up the country from Maryland, from the Head of Elk – after the British crossed Christine Creek they had an engagement with the Americans at a place called "Bunkers Hill" in Delaware State, but deponent was not in it – From Wilmington we then marched towards Brandywine where we had a battle, our brigade was not in the battle, General Wayne's division was engaged, deponent did not see Washington there but he was there, though not in the battle – From Brandywine deponent was marched to The Trap about 27 miles from Philadelphia in Montgomery County Pennsylvania. Washington, General Potter, General Green, and General Wayne were at The Trap. From there we were marched to Stony Run, there we got orders in the morning to clean our arms and load them – In the morning when the roll was called we had to fall into rank and march in the night down to Germantown, we marched the whole night and in the morning about day break Washington's cannon's fired and the battle began – All the generals who were at The Trap were at this battle, General Stevens was also there and commanded at Frankford Road – The battle lasted until about 10'0clock when our men had to retreat after the British came from Philadelphia – From Germantown we marched towards Hickorytown and crossed the Schuylkill at Swedes Ford and marched to Valley Forge and lay there until deponent was discharged, which was the 26th day of December 1777 – General Washington was with us at the Valley Forge.

Author's note – Frederick Kalehoft's pension application was approved and he received $55.52 per month for 16 months and 20 days of service as a private.

There is no mention of his date of death.

John Kessler

In this very interesting application John Kessler covers a few very interesting points about the earlier pension acts. He also mentions several very important people of that period during his time of service aboard the Frigate Alliance.

As I have previously mentioned, there were several pension acts before the 1832 pension act, the one that you read about most often in this book. John Kessler is a very unusual pensioner in that he filed in 1794, under the original invalid pension act. Just about all invalid pensioners to file under the initial pension act had passed away by the 1832 act. Under the initial invalid act, he was receiving 1-2/3 dollars a month until 1816, when it was raised to 2-2/3 dollars. During his application for the 1832 pension act, which was more than likely for more money, he encounters difficulties because his paperwork was burnt in one of the War Office fires in the early 1800's.

Serving aboard the Frigate Alliance he mentions transporting such dignitaries as Thomas Paine, the famous writer of his time. Alliance was a very fast and well-armed 36 gun frigate and one of her duties was to carry dignitaries, money, arms, and dispatches back and forth to France.

John Kessler joined the Alliance in Boston while she was undergoing repairs. He arrived at a turbulent time in Alliances' history when her captain up until that point, Captain Pierre Landais, was relieved by Captain John Berry.

State of Pennsylvania County of Philadelphia. On this fifth day of December 1836 personally appeared before the Recorders Court, for the incorporate district of the northern liberties, and the districts of Spring Garden and Kensington in the County of Philadelphia and State of Pennsylvania: in open court; John Kessler a resident of the Northern Liberties in the county and state aforesaid aged 75 years who being first duly sworn according to Law doth on his oath make the following declaration in order to obtain the benefit of the provision made by the Act of Congress passed June 7th 1832.

That on the 28th day of November in the year 1780 he entered as a midshipman on board the Continental Frigate Alliance, John Berry Esq. Commander, then lying at Boston in the state of Massachusetts. That he served as a Midshipman on bard said frigate to the 9th of December 1782 and from which time he served as a Master's Mate to the conclusion of the Revolutionary War and was discharged from said frigate then at Providence in the state of Rhode Island in the month of April 1783. That for injury received by him while acting as a Midshipman aforesaid he in the year 1794 applied for a pension and having produced the requisite vouchers and proofs of his service as a Midshipman aforesaid and the injury *received he was placed on the Pension Roll of the Pennsylvania Agency at twenty dollars per annum and which he received until the Act of Congress passed the 16th of April 1816, his said pension was raised to thirty two dollars per annum and which last amount he has received to the 4th of September last. That the papers upon which he was pensioned as aforesaid (it is said) were destroyed either in the year 1801 or 1814 when the War Office was burnt and it is now required of him to produce new vouchers of his service and statin on board the said frigate before he can receive the benefit of the provision made by Act of Congress passed the 7th of June 1832.*

The declarant says that he has no knowledge of any person living who has personal knowledge of the declarant's service and station on board the said frigate. That the best proof which the declarant can produce are contained in the service papers hereunto attached, three of which are signed by John Berry and the handwriting of said signatures verified by the deposition of Patrick Hayes hereunto annexed and who also deposeth that the name of the declarant appears in the books and papers of said Barry's estate as a Midshipman on board the said frigate Alliance.

That the declarant has seen a report made by the Honorable John Armstrong, late Secretary at War on the 31st of May 1813 to the House of Representatives relative to the Invalid Pensioners of the United States in which report the name of the declarant as a Midshipman doth appear. That during the service of the declarant as aforesaid the said frigate was variously employed, she sailed from Boston on the 11th of January 1781 conveying Colonel Lawrence (the younger) on an embassy from Congress and also Major Jackson, Secretary *of Legation, Thomas Paine the noted political writer in the Revolution was also a passenger on board. She arrived at L'Orient on the 10th of March following. On her return passage she captured two British Privateers, two Merchantmen, and on the 28th of May two Sloops of War and with which last two she had an action of more than 2 hours continuance and in which action Captain Berry was severely wounded among the crew eleven were killed and a number wounded.*

Author's note – Alliance had engaged the British sloop's Atalanta and Trepassey. In the action Captain Berry was struck in the shoulder by grape shot.

On the frigates arrival at Boston Captain Berry dispatched the declarant to Philadelphia for Mrs. Berry and on which accession he gave the declarant the paper hereunto attached dated at Boston June 12th 1781 certifying the declarant to be a midshipman on board the Alliance and to be passed unmolested to Philadelphia. The Alliance wanting very considerable repairs remained in Boston Harbor until the 25th of December 1781 when she sailed having the Marquis de La Fayette on board and who was to be conveyed with all possible dispatch to France and arrived at L'Orient on the 18th of January 1782. After waiting some time for dispatches from Dr. Franklin to Congress she again sailed and arrived at New London in the state of Connecticut on the 13th day of May 1782 and on the 4th of August following she sailed on a cruise during which she captured seven vessels, four of which were heavily laden and bound from Jamaica to Glasgow. They were all ordered to proceed to L'Orient in France and arrived there in company with the frigate on the 18th of October 1782.

The declarant being Master of one of the said prizes received from Captain Barry the instructions contained in the paper attached hereto dated at sea on board the Alliance 28th September 1782 and also another paper purporting to be the private signals between the said Barry and the declarant in the case of their separating and meeting again. On the 9th if December 1782 the Alliance sailed from L'Orient again on a cruise during which she went into the harbor of St. Pierre in the island of Martinique where Captain Barry found orders proceeding to the Havana to take in Specie for the United States and accordingly she sailed to the Havana and after taking a large quantity of Specie on board she in company with the Continental ship Luzerne of about twenty guns sailed for Philadelphia.

Author's note – Specie is coin money.

On the passage three British frigates gave chase, one of which Captain Barry permitted to come up and engage for about 45 minutes when the enemy's ship being very much shattered declined any further contest and joined her consorts made sail from Alliance in consequence of a French 50 gun ship being in sight. On arriving at the Capes of Delaware a British ship of sixty-four guns gave chase and compelled the Alliance to leave the Capes and on the 25th of March 1783 she arrived at Providence in the state of Rhode Island and where the declarant and the crew were discharged from the frigate (Peace being proclaimed).

The foregoing facts are all fresh in the memory of the declarant but for most of the dates of their occurrence he is indebted to the remains of the journals and notes kept by him while on board said frigate. The declarant is 75 years of age and a native of the city of Philadelphia and that in the years 1779 and 1780 he went several voyages from Philadelphia to the West Indies, on the last of which he was captured by the British and carried to Jamaica from where he made his escape and got to Boston where he entered the frigate Alliance.

Author's note – John Kessler's pension application was approved and he received $150.00 per year for 20 months as a midshipman and 4 months as a Mate.

He passed away March 17th 1840.

James Knight

This is another claim that includes both land and sea service, but is the first one I have come across that mentions service as a Marine. His service aboard the frigate Randolph is once again a nice document of naval history. Mr. Knight glosses over his service but I have added "author's notes" to expand on this important part of naval history. And as you read my final notes on this chapter you will see that James Knight is a lucky man.

State of Illinois, Edgar County. On this twenty seventh day of September 1832 personally appeared in open court being a court of record to the Circuit Court in and for the said county of Edgar James Knight, a resident of the county of Edgar and state of Illinois aforesaid aged 82 years the 20th August last who being duly sworn according to law doth on his oath make the following declaration in order to obtain the benefit of the provision made by the act of Congress passed June 7 1832.

He states that he was born in the county of Philadelphia on the 20th of August 1750, this his age is given from the register in his father's family bible. That about the first day of July 1775 *he enlisted in the service for twelve months in the county of Bedford and state of Pennsylvania under Robert Cluggage, captain in Magaw's battalion in the Pennsylvania regiment of rifle men commanded by Colonel Thomson, in a few days was marched to Carlisle in Cumberland county in said last named state, thence to Eastown is said state, thence to headquarters at Cambridge in the state of Massachusetts where we arrived about the last of August or first of September and encamped there till the 13th of March following. Then marched to the city of New York and after two weeks marched to Long Island and there got my discharge for my full term of 12 months service as aforesaid, which said discharge is lost. That about two weeks he again enlisted as a marine under Captain Shaw for 12 months and entered on board the Frigate Randolph of 32 guns, commanded by Captain Nicholas Biddle, William Barnes 1st Lieutenant, Thomas Douglass 2nd Lieutenant, and Joshua Fannin 3rd Lieutenant, and lay at Mud Island below Philadelphia until the 5th of February 1777 when said Frigate put to sea on a cruising voyage being about six weeks out put into Charleston in the state of South Carolina.*

Author's note – The Randolph was commissioned in July 1776, so James Knight was on it commissioning crew. The cruise he mentions is her maiden voyage which escorted American merchant vessels out to sea. As she made her way to Charleston fever broke out aboard the vessel and a great number of the crew succumbed to the illness and were buried at sea.[6]

And while here his said second term of 12 months enlistment expired, but the captain would not give him a discharge, saying we must take another cruise and try to do something as we had done nothing yet. And there we lay until August in which time our main mast was twice struck with lightening and later repaired and the last time a conductor was placed leading from the mast head into the water. Sometime in August aforesaid set sail for another cruise and in 3 or 4 days fell in with a British ship of 20 guns, one of 12 guns, two brigs and a sloop of 8 guns all belonging to the British.

Author's note – The 20-gun vessel was the True Briton, laden with rum. The 8-gun sloop was the Severn, and laden with rum, sugar, ginger, and logwood. The two brigs were the Charming Peggy and the L'Assomption, both laden with salt. The 12-gun vessel is unknown.

With which said vessels we had an engagement and captured four of them, the sloop escaping. We brought them into Charleston, one Brig was discovered to belong to the French and was given up to them, the other 3 and cargo was sold under the prize law, the proceeds of which sales amounted to £658.15 to each private or soldier. The Randolph lay in Charleston till 15 December following when this applicant got his discharge, which is lost, and in a few days started home to Bedford County in the state of Pennsylvania where he arrived sometime in the month of March 1778. After 2 or 3 weeks he again enlisted for 9 months under Thomas Cluggage, captain, in Major Robert Cluggage's battalion. Marched to Sinking Spring Valley where we built a fort called the Lead Mine Fort and continued in service there till his said last 9 months term of service expired. He then continued in service as a volunteer for 12 months longer and served the whole of his last mentioned volunteer service and left the service in the year 1779.

Author's note – James Knight's pension application was approved and he received $40.00 per year for 2 years of service. One year as a private in the army and one year as a private in the marines.

He passed away February the 23rd 1838.

The Frigate Randolph was lost on the 7th of March 1778 when she went up against the much bigger 64-gun HMS Yarmouth. The Randolph's magazine was hit and she exploded killing 311 of her crew. Only 4 survived the blast.[6]

Thomas Laidley

In another claim that covers multiple services Thomas Laidley's application describes his tour in the army, navy, and as a steward in a hospital. With all his service, he is for some reason only credited for 10 months in the navy as a gunner - although he had much more service, even receiving a captaincy in the navy. It's a very interesting claim that covers a lot of ground and numerous battles.

Virginia Cabell County Court. At a Court hereto for Cabell County at the court house on the 28th day of October 1833.

On this twenty eighth day of October 1833 personally appeared in open court before the county circuit court of Cabell County now sitting Thomas Laidley a resident of Cabell County aged seventy eight years the 1st day of January 1833 who being first duly sworn according to law, doth on his oath make the following declaration in order to obtain the benefit of the act of Congress passed June 7th 1833.

That he enlisted himself as a private soldier in the state line of Pennsylvania for the tour of three years from the date of his enlistment in March 1776; in the first independent rifle company of Pennsylvania; John Doyle Captain, Samuel Brady First Lieutenant – That he continued in the same service in the capacity of a private soldier in that company a year.

In March 1777 this applicant enlisted one Michael McDoniel to serve in the same company, in his place during the war, and this enlistment discharged this applicant from the service of the remainder of his term of enlistment and he was there upon discharged.

This applicant then immediately, he believes in the month of March 1777 enlisted under Captain William Lisle (or Lyle) commander of the lookout boat "Resolution" of the Pennsylvania fleet under Com. John Hazlewood then lying at Fort Mifflin in the Delaware and served in the capacity of Gunner.

Shortly after this applicant entered the navy, he believes within the space of two months, he was favored by Com. Hazlewood with a captains commission, and the command of the boat "Resolution." Capt. Lisle (or Lyle) having been transferred to one of the gallies – This applicant continued in the service and in command of this boat until the British took Fort Mifflin and Red Banks in the month of November 1777. In the month of May 1777 Com. Hazlewood ordered this applicant with his boat and four or five other boats to take charge of the fire rafts (rough boats filled with combustible materials) to ascend the river, and endeavor to burn the British. This applicant believes the Roe – Buck then at anchor some distance below Fort

Mifflin. He obeyed and crossed down as far as Reeds Island, but the British slipped their berth and put out to the Bay.

This applicant was in an engagement in defense of the fort at Red Banks opposite the mouth of Schuylkill when his, together with other armed vessels rendered some services which fort was attacked by the Hessians under command of General Dunap and another the name cannot recollect, the enemy was repulsed – The two generals wounded and taken prisoners (Dunap dying soon after of his wounds) and this applicant aided in putting them on a boat, in which they were sent to the British then in Philadelphia.

Author's note – He is speaking of General Carl Donop who was wounded and within days, died of that wound.

When the Battle of Brandywine was expected to be fought early in September 1777 this applicant with his boat together with several other vessels under the command of their respective captains (this applicant thinks Captain Martin, Lyons nd Wilkins were among them) were ordered to Chester about twelve miles below Fort Mifflin to aid the American army to cross the Delaware in case it should retreat.

After the Battle of Germantown, the British being in possession of Province Island (below the mouth of Schuylkill at Watts establishment) cut a port hole through a bank which was thrown up on the island some years before for the purpose of reclaiming the soil – opposite the block house on Mud Island – an island proper in the Delaware River – and placed an eighteen pound cannon and fired upon the block house.

Captain Henry Martin of one of the boats applied and obtained leave of Com. Hazlewood to man a boat carrying a four pounder and to approach so near as to be able to fire in this port hole. This applicant, Captain Lysle, and perhaps one or two more captains, together with several privates, volunteered and executed the desperate attempt under a heavy shower of grape shot and returned without any injury.

One or two days afterwards, the whole of the British soldiers, except perhaps the principal officer, amounting to forty or fifty stationed at this port hole deserted and came down opposite Fort Mifflin hoisting a white flag.

This applicant and Captain Martin with their boats were ordered to go over and receive them – which they did – bringing them to the fort, and they were afterwards sent off to the Jersey shore – at Fort Mifflin the artillery was commanded by Col. Thomas Proctor, the infantry by Col. Smith, and as stated as herein before the fleet by Com. Hazlewood

The Chevaux de frise placed in the main channel of the Delaware with the drift collected thereon had diverted the water principally to the Pennsylvania shore. This applicant, Captain Murphy, and one or two other captains were ordered to sound this channel when they found it had increased very much in depth. That night Captain Murphy with his boat's crew deserted and went up to Philadelphia; from the information by him communicated, the British sent off to New York – sent down a three decker, the Vigilant, mounted navy cannon, brought her around and anchored in this channel between Mud Island and Hog Island and by this means effected the destruction of Fort Mifflin on the 6th of November 1777.

Captain Martin (or Be Jesus as he was familiarly called), this applicant, and several others volunteered to burn the British frigate Augusta, a ship of war which had run aground at the mouth of Mantua Creek, a desperate undertaking, but before they reached her she blew up.

After the fall of Fort Mifflin and Red Banks this applicant was ordered to Trenton in New Jersey in the month of November 1777. This applicant was advised that as no allowance was made for services as steward of the hospital of the fleet it is unnecessary to specify his services in that capacity further than to state that he served about one year, holding during that time his commission of captaincy and himself subject to the command of his Commodore Hazlewood whenever his services as captain should be required. During part of that year several of his fellow captains who like himself were under commission, but out of service, drew their rations and their pay regularly as he did.

In November 1778 his health being bad he was permitted to withdraw from the service.

Author's note – Thomas Laidley's pension application was approved and he received $36.00 per year for 10 months of service as a gunner.

He passed away March the 17th 1838.

Ephraim Lewis

This pension claim caught my interest because it's the first I have read that describes Colonists wearing British uniforms in disguise. It also has an interesting twist to it and a sort of foreshadow of what is to come with General Benedict Arnold at West Point a few years later, when he is discovered as a traitor.

The practice of hanging and punishing Tories was common. And in just the previous pension application by Abraham Lewis, he stated how they caught several Tories suppling provisions to the British. He stated that *"He assisted in hanging six Tories, two at one time and 4 at another and let two off by their taking five hundred lashes well laid on and offering to serve in said company so long as its services should be required. The Tories that was executed was left hanging on trees on the bank of the Delaware River."*

Ephraim Lewis gave the following statement for his pension application;

State of Tennessee, Carter County. On this 16th day of June 1840 personally appeared before Isaac Tiptan one of the justices of the peace for said county Ephraim Lewis a resident of Carter County in said state aged eighty two years deposeth and sayeth in his the following declaration in order to obtain the benefit of the act of Congress passed June 7th 1832 that he entered the service under the following officers and served as here after stated. That he volunteered in Buck County Pennsylvania the thinks in the year 1777 and served three months under Captain John Wisner, Col. McCambley's regiment and marched to West Point Fort and was employed in keeping garrison an at the expiration of three months or thereabouts was verbally discharged. Very shortly after returning home he thinks in the same year 1777 he again volunteering from Buck County Pen. under Captain Welling and was marched and served three months the time he had volunteered to serve under Col. Stethhorn, in this service was attacked by the British Light Horse and after a short engagement we drove them and kept the field and again we received verbal discharge. And very shortly afterwards he thinks in the following year 1778 he volunteered from Susquehanna River in Lehigh County Pen. under Captain David McCambly and was marched to West Point Fort and there joined General Lincoln and was marched to meet the British on their way to a town called Boskos and arrived after the British had found the town but were still aboard one of their vessels, we encamped there the night. The next morning Lieutenant Luckey with twelve soldiers of whom deponent was one was ordered out to reconnoiter the town and observe and detect spies and enemy if any, our party being disguised in British uniform and while out a good looking man came up and gave his hand to the lieutenant and accosted him saying he was glad to see him there and when attempting to withdraw his hand the lieutenant held him fast telling him he was his prisoner. The man instantaneously drew something from his pocket and slipping it in his mouth as we supposed he swallowed something, we instantly marched him to camp and presented his case to the general. Medical aid was immediately called in and after having given him large doses of purgatives, some kind of a metallic ball passed from the prisoner, deponent believed to be silver in which was found a letter from Lord Cornwallis to Burgoin with a request forthwith to march his forces and join him in Philadelphia Pen. After trial and consultation the prisoner was hanged on an apple tree in the presents of the deponent immediately afterwards the letter was dispatched to General Washington and we were immediately ordered to go join General Arnold and we arrived during the engagement between Arnold and Burgoin. As we advanced Arnolds troops opened to the right and to the left and we were marched in the center, after one fire we charged bayonets and deponent was an eye witness to the surrender of General Burgoyne and we were then marched back to West Point Fort having charge of the prisoners taken in the engagement.

Author's note – Lewis Ephraim is describing the Battle of Saratoga and the surrender of General Burgoyne entire army on October 17th 1777.

And after having served the term for which we volunteered we were again verbally discharged as deponent believes shortly thereafter perhaps in the next year deponent volunteered again from Bucks County Pen. under Captain Jetadine Gour, commanded by Col. Dennisan and marched to a place calle

Forty Fort and joined Col. Butler, a continental officer who took the command. Some distance from where we were stationed the Indians and Tories had taken possession of Hunter Mills Fort. We were marched to dislodge them. The enemy met us on the way and after an engagement our forces were defeated with a heavy loss, our retreat was in confusion, every man making for himself – Myself with some others swam the Susquehanna River twice in our retreat and arrived at the Fort some time in the night, after which we abandoned the Fort and county for which the enemy having overcome us. After some time we were ordered to return and did so, under Col. Butler. Deponent then served two summers as the commander of a small scouting party under the immediate command of Col. Butler.

Author's note – Ephraim Lewis' pension application was approved and he received $20.00 per year for 6 months of service as a private.

He passed away October the 8th 1845.

William Lyons

This is once again a story of both land and sea service, but what makes William Lyons story interesting enough to make this book, is that he served even after suffering a significant wound.

The State of Ohio, Morgan County.

On this eighteenth day of July in the year of our Lord One Thousand Eight Hundred and Thirty Two – Personally appeared in open Court before the Court of Common Pleas for said County in the Eighth Circuit of the State of Ohio now sitting, William Lyons a resident of Olive Township of said Morgan County and State of Ohio aged seventy five years being duly sworn according to Law doth on his oath make the following Declaration in order to obtain the benefit of the act of Congress passed June 7th A.D. 1832.

That he entered the service of the United States under the following named officers and served as herein stated, that in August in the year of our Lord One Thousand Seven Hundred and Seventy Six, he the said William Lyons entered the Regiment of Pennsylvania Militia commanded by Colonel James Crawford & Lieutenant Colonel Samuel Patterson in a company commanded by Captain John Scott – First Lieutenant William Smith, Second Lieutenant John Underwood & Ensign William Gwinn in Leacock Township, Lancaster County in the State of Pennsylvania for the term of two months as a drafted Militia man, from whence they marched to Philadelphia, from thence to Perth Amboy, from Perth Amboy to Bergen, at Bergen the tour of two months expired and we volunteered for six month into a company commanded by Captain William Scott – First Lieutenant James Armer, Second Lieutenant James Hughston, Ensign James Woods – in the Second Battalion of the Lancaster County Flying Camp commanded by Colonel Clotts, Lieutenant Colonel Thomas Morrow, Major John Boyd – From Bergen the regiment marched to Paulus Hook on the day that the English took New York – On the evening of the same day the Regiment returned to Bergen – from thence to Perth Amboy where Colonel Clotts had remained from Perth Amboy a part of the Regiment was advanced to Staten Island to attack a place called Cockles town, including the said William Lyons, in taking the town a small skirmish ensued in which he was engaged and they took Hessian prisoner, the first that had been captured by the Americans. From Perth Amboy the Regiment marched to Fort Lee where it remained until the month of November – In November one hundred and thirty seven men were required of Colonel Cotts Regiment to go to Fort Washington on York Island – the was raised by volunteers of which the said Wlliam Lyons was one – For which place they marched commanded by Lieutenant Colonel Thomas Morrow to Fort Washington and on either the sixteenth or seventeenth of November a battle was fought in which the said William Lyons received a musket ball in his right thigh, which remains in his thigh at the present time. He was taken back to Fort Lee to the hospital where he remained five or six days, from thence he was removed to the hospital at Morristown t Bethlehem in the State of Pennsylvania where he was dismissed about the latter part of January by John Hughston Surgeon, the time for which he had volunteered having nearly expired but which discharge has been lost or destroyed. Being unable to serve as formerly because of the wound he had received he went in the year Seventeen Hundred & Seventy Seven to West Fallen Field in Chester County Pennsylvania to learn the shoe making business – when he was drafted and served another tour of two months in a company of militia commanded by Captain Gibbs – in a Regiment commanded by Colonel John Hannams & Major Hartman, Was dismissed at Billingsport – In September of the year Seventeen Hundred and Seventy Nine he volunteered in Philadelphia and went aboard the ship Flora of twenty guns commanded by Captain Samuel Downs. From thence they sailed for the West Indies and returned to Philadelphia in July Seventeen Hundred and Eighty, after a voyage of about ten months where he was dismissed.

Author's note – William Lyons' pension application was approved and he received $66.66 per year for 1 year and 8 months of service.

There is no mention of his date of death.

Peter Mathews

You don't hear the term fifer in modern military terminology, but during the Revolutionary War they were an important part of the military organization. They, along with the drummers, sounded signals for formation changes. These positions were usually filled by young boys, hence his young age of 13 years old when he joined military service. Not only is Peter Mathews application notable as a young fifer, but his is the first I have come across that states that he was transported to England as a prisoner of war.

State of Kentucky, County of Hopkins.

On the tenth day of February One Thousand Eight Hundred and Thirty Five personally appeared in open court before the justice of the court for the County aforesaid now sitting Peter Mathews a resident aforesaid County stated age seventy years who being first duly sworn according to law doth on his oath make the following declaration in order to obtain the benefit of the act of Congress passed 7th of June 1832.

That he entered the service of the United States under the following named officers and served as herein stated – That he has no record of his age but from the best information which he derived from – his parents, he was born in what was then Philadelphia (now Montgomery) County in the state of Pennsylvania in the month of July in the year 1763. And in that county at the residence of his father where he had resided from birth – in July or August 1776 he volunteered as a fifer for a tour of two months under Captain Kegger and Colonel Bird of the Pennsylvania Militia that had rendezvoused at his residence aforesaid and marched thence to Philadelphia where he remained a few days and then marched through Trenton in New Jersey, thence to Princeton, Brunswick, Elizabeth Town, to Newark, to Paulus Hook where he was stationed until his term expired when he was marched to Amboy and discharged – he received a written discharged signed as he believes by his captain, but has lost the same. He further states that at Amboy aforesaid immediately after he was discharged he again volunteered for a tour of six months under Captain Redheffer of German Town Pennsylvania of the Pennsylvania Militia and was in a regiment commanded by Colonel Smith in which Bush was a major – Shortly after he volunteered he aided in capturing some English and Hessian soldiers on Staten Island, this surprise was headed by Colonel Smith who was wounded in one of his arms.

From Staten Island this applicant was marched to Amboy and from thence to the neighborhood of Fort Lee and was stationed there when General Washington reached that post from the White Plains with the main army. At this point, about one month after he volunteered under Captain Redheffer, this applicant and the picket guard with which he served was surprised by parts of the enemy Light Horse and this applicant was wounded in his left leg and taken prisoner and was conveyed to the city of New York where the enemy at that time was stationed under General Sir William Howe.

That after remaining at New York a short time this applicant was placed on board of a British prison ship and transported to Hallifax and from thence to Liverpool in England where he was quartered in the house of one James Brindle and detained until the spring of 1781 when he was permitted to embark as a hand on board of a British merchantman bound to the West Indies and when out at sea eight or ten days the merchantman was captured by an American Privateer and taken into Boston where this applicant was released and returned home having served his country faithfully in the field being wounded and having while yet a boy suffered an awful captivity of more than four years in a land far from his friends and his country.

This applicant is unable to account for his not having been exchanged otherwise than by that presumption that he had been reported as slain at the time he was taken prisoner. He made several efforts to inform his friends of his situation all of which failed. He being young as well as illiterate in all probability he did not use the means least calculated to affect the object.

Author's note – Peter Mathews pension application was approved and he received $33.33 per year for 6 months as a private and 2 months as a fifer.

He passed away the 13th of November 1840.

Angus McCoy

This is another finely detailed application which describes service on the western frontier. Angus McCoy's detailed account of fighting at Crawford's Defeat is amazing and you feel as if you are right there with him. He is a well-spoken man with a broad vocabulary. The end of his application has a familiar tone that I find in a majority of the applications I read. A pride that is tugging on them not to apply for a pension, but age and physical disabilities that force them to have to ask for assistance.

Sworn in Open Court September 23rd 1833

Your petitioner Angus McCoy in the seventy third year of his age and after March next, will commence his seventy fourth year do make the following declaration in order to obtain the benefit of the act of Congress passed June 7th 1832.

On this first day of January 1833 personally appeared before David Frazier a justice of the peace in West Finly Township in the County of Washington in the State of Pennsylvania who being duly sworn according to law doth make the following declaration in order to obtain the benefits aforesaid.

I Angus McCoy according to my mother's statement was born March 1760 in Scotland. We landed at Philadelphia in the fall of 1772, I being upwards of twelve years old (my father dying in Scotland), after I arrived to twenty one I moved to the western part of Pennsylvania in Washington County and Chartiers Township, and there in the summer of 1781, the Indians being very troublesome on the then frontiers or new settlements. I having no family at the time volunteered to guard the then frontiers on Chartiers Creek and now Chartiers Township during the time of cutting and gathering in the then harvest, in their collective capacity from field to field and house to house forting to gather at night and laying with our fire arms in our arms occasionally – In the same summer after the before mentioned service it was rummered that three hundred Indians were in hostile array at the Moravian towns, by a squaw that came to Fort Pitt (now Pittsburgh) at which report Captain Van Swearingin raised a company of volunteers of which I was one. We marched to Robinson's Run in this county where we had to detain until we procured ammunition from Pittsburgh. We then took up our line of march and proceeded by stream and ridges having no road until we struck the Ohio River at or near the place where Georgetown now stands. There was three of our men went into the river to swim or bath themselves. They discovered the Indians about a mile below us, making over to the other side. We ran down with all the might or speed we could, they discovered us. They sprang off the horses which they had stolen from our side, they principally swam back. But the Indians made their escape, we proceeded about ten miles down the river bottom to opposite Yellow Creek and returned up Harmons Creek, a circuitous rout making no other discovery of Indians – The time I lost as a volunteer this summer in first guarding during harvest time and the above expedition my memory will not serve me – The last expedition we traveled some times the meanderings of streams and ridges – I find that the direct road at present exceeds a hundred miles circuitous – The committee on this petition will please allow me what time they think proper on the above and following services

In the following year 1782 early I the spring, I was drafted a tour of military duty according to the then Pennsylvania law and served it faithfully under Captain Andrew Swearingin at Burgets Station at or near where Burgetstown now stands guarding the then frontiers by traveling from point to point and house to house and field to field while the people were working in their collective capacity – Immediately at the close of the above draft I volunteered and served a tour of military duty in the room or place of my brother William McCoy who had a charge of a family and was legally drafted next in rotation to the one I had just served. I have no documentary evidence in my possession by discharge if any was given for these two drafts and I do not know of any of my comrades living and shall therefore give you a detailed account of the last above volunteer service in the room of my brother who was drafted in succession of my former draft according to the then Pennsylvania law (this tour was of such a matter as to leave a lasting impression) I was placed under Captain Charles Bilderback a noble officer who was under the command of Colonel Crawford – We each found our own horses, firearms, and equipment – we crossed the Ohio River at the Mingo Bottom – we there detained until ammunition was procured from Pittsburgh. Our number being four hundred and eighty nine, our distance was computed by our pilots who were Meyers, Nicholson, Slover, and Zane to be two hundred miles without any road. We proceeded thus through the woods and crossed the Lusconawrus waters at length we reached the Sandusky plains. It was about midday when we entered – we found a path which our guides knew led to their towns, we traveled until night, and camped as usual. On the morning of the fourth of June (for on this day we had our battle) we took up our line of march and having come to the place where their town formerly stood, it was removed except the back walls of their fire places, we being by computation then thirty miles in the plains – Here we sent our spies ahead who had not gone far before they discovered the Indians. The Indians gave chase to our spies and ran them hard. (William Midkirk one of the spies) having a swift horse soon appeared giving us the signal, we each mounted and went to meet them as fast as we could, there was one Indian who did out run his comrades to meet us. But he was the first killed. We dismounted and endeavored to screen our selves behind the few remaining trees and let them advance on us. The play of humane destruction began – I think it was a little after the middle of the day when this louder music began. We continued watching and firing at our adversaries and them at us until dark occasionally. Soon after the commencement of the battle, A number of our men got wounded, some badly and some fell to rise no more – I got my own clothes riddled with balls but a merciful Providence preserved my flesh – Finding that the Indians were concealed in the long grass, Daniel Leet mounted his horse and as he *passed me looked me in the face said follow me. I immediately gave the same invitation to those around me who were on foot. I took after Leet who rode between a cantor and a gallop and I suppose between fifteen and twenty after me. We routed them in groups out of the grass. In this daring maneuver in their consternation not a gun was fired at us until Leet wheeled to the left at which time two Indians discharged at Leet. I saw his horse bounce as if mortally wounded (but neither injured). Our pass was so quick we had no time to fire on them, and a kind Providence prevented them. We supposed that we passed at least one half of the Indian line. The reason why Leet I suppose selected me was we were well acquainted having served on a former draft together (but he is a few years past numbered with the dead or I have not the least doubt would corroborate this statement) Shortly after this a brother soldier close by me got wounded. I asked him for the loan of his gun, it being a superior one to that of mine. He gave it to me and his ammunition. (Mine I left on the battle ground) I confess I felt my self stouter being prepared with my additional stock of ammunition. We had some as brave a man as was shot a gun, we had some it is true were no credit to themselves but they were the fewest number. I think we were at this work of destruction for at least five hours, dark at length prevented us, the night being short at this season of the year (being the 4th of June). We were all anxiety. And expected that the same course would be pursued in the morning – But the Indians did not advance on us, and our officers gave us no order to advance on them – During all the night past and all the 5th of June they were reinforcing themselves and forming a circle around us. Keeping at a distance a while before sun down there was a company of Indians coming up to their place in the circle who* *discharged their pieces at the sun – near or about this time they had completed their circle and began to show themselves except where there was a deep swamp. Here they had placed a sentry and nearly central in this swamp – We supposed their intention was when all their forces were collected to close on us at or after dark. We however got everything in readiness, our wounded on horseback. (through the day we buried our dead and burned what stuff we could gather over where they were buried in order to deceive and prevent them from raising them)(we left non on the ground but one man who was shot through the breast who could not live long by the name of Thomas Ogle). In this critical situation after sun down about the closing of the day we made an attempt to retreat at which time the sentry in the center of the swamp discharged his piece. And immediately all in the circle discharged their pieces we supposed at us. But without effect. Which proved very favorable to us for before they could reload we passed them principally thru their lines and some thru the swamp. Some of which stuck fast and fell prey to the enemy – We marched all night as fast as the wounded could bear and circumstances permit – The Indians did not annoy us any more that night. We supposed from taking a wrong road or path from that intended. The next day we made across through the plain and continued our course as fast as our wounded could bear. Keeping together as well as we could in the evening some of our small army being a small distance ahead. The Indians laying in ambush, rushed on them and caught John Hayes and before we could rescue him they had his scalp half raised off his head and inflicted a mortal wound with a tomahawk on the same. And shortly after and before we were out of the plains. A body of Indians both on horse back and foot attacked our rear – our* small army gave them powder and ball in exchange – While our front gained the woods with the wounded. Some of our men were here killed and others wounded. John McDonald got his thigh bone broken with an Indian ball or slug – Captain Bilderback requested me to take charge of McDonald. We got to the wood about dark and there we camped for the night. It commenced and poured down rain and that very heavy for the fore part of the night Hayes, who was still living and McDonald was laid on one blanket. My business was to guard them. I walked around them all night with the lock of my gun under my arm in order to keep it dry. At the break of day things being in readiness we started, I got McDonald on a horse. (Hayes was still living but could not live long we left him there) The Indians did not attack us anymore – I brought McDonald home with much difficulty having to lead his horse all the way as we had no road – He died in a few days after arriving at home – I believe all the rest of the wounded were able to guide their own horses – While I am on this sickening campaign not wishing to wound or hurt the feelings of any connected or concerned. I understand that Colonel Crawford the first night of our retreat left our little army taking with him Slaver the pilot, Harrison his son-in-law and Doctor Night – One thing I know and am certain of is I never saw Colonel Crawford after the fore part of the first night of our retreat – I understand they were taken by another party of Indians who were coming to assist in the engagement – Colonel Crawford I understood was burnt to death – Slover I saw some years afterward. He I understood was tied to a stack or tree to be burnt, but there coming on a great rain the Indians supposing that their hellish pleasure would not be gratified, loosed him for that time as he supposed for a better evening, during which time he made his escape – Doctor Night as miraculously made his escape but had nearly perished with hunger – I never learnt what became of Harrison.

In the same fall after Crawford's Defeat I was again drafted, our frontiers being few in population we had to be one half of our time on public duty. Shortly after my time had expired wherein I had volunteered to serve in place of my brother before stated, and when on our way to Fort Mackintosh our orders were countermanded – Here I cannot recollect the time lost. I know it was late in the fall and I was glad our orders were countermanded for I dreaded the winter campaign in our unprepared state – One thing I know that from early in the spring to late in fall of 1782 my principle business is above stated, our difficulties of fatigue, hunger and uneasy situation of mind is bright in my recollection. But the exact time I served I cannot recollect. Yet I feel fully persuaded that what ever time the Pennsylvania drafts were for I served for myself and for my brother as before stated – I shall here state that I never received anything for my military service other than certificates, which I traded for little or no value to the best of my recollection. I have never received any pension. I neve was an applicant until at present for a pension. And now I only claim your lowest grade of pension, which from the former and after services rendered by your petitioner will I hope appear sufficiently satisfactory to your committee to place me there on – I am at present living in Washington County Pennsylvania about thirteen miles from Washington and have been for upwards of thirty years. I am a resident od said county for fifty one years – I am nearly confined to my residence. I have been disabled by rheumatic pain in my left loin or sciatic joint for *nearly twenty years – Immediately after peace was in some manner restored in some degree to the neighborhood of Washington – I being a single man and having no charge I moved to the then frontiers bout thirty miles north west of Washington – Where I volunteered on several short scouts sometimes in a position of a ranger where I guarded, forted, and worked with the people in their associate capacity for a summer season – I next moved to Wheeling waters where I continued a few years until I settled where I now live, 16 or 17 miles west of Washington where the Indians had formally been very troublesome – After my arrival there, two young men by the name of Beham was taken prisoner within two miles of me – Likewise one man and women was killed within a few miles of me. As late as 1791 I have laid down at night with my gun in my arms to defend myself from the before mentioned natives.*

Author's note – Angus McCoy's pension application was approved and he received $20 per year for 6 months as a private.

There is no mention of his death.

It's interesting to note that with such a detailed pension application, his initial pension claim was rejected by the Pension Board. They stated that in their estimation Angus McCoy only qualified for 5 months of service, and needed 6 months to obtain a pension. A subsequent letter was drafted by Angus McCoy's lawyer which detailed dates and times he served. The Pension Board satisfied with this new information awarded him a pension for 6 months of service as a private.

James Mc Kinzey

Although this is a very short pension claim, I have included it because Mr. Mc Kinzey sailed on four of America's first war ships. If this feat isn't interesting enough, he sailed on three of the ships with only one leg.

It's also a unique claim because it is from 1818, which are usually very limited in information.

District of Pennsylvania

On the third day of April A.D. 1818, before me Richard Peters, Judge of the District Court of the United States, in and for the Pennsylvania District. Personally appeared James Mc Kinzey, who being duly sworn, deposeth and declareth, that during the war of the Revolution he served against the common enemy as a seaman and afterwards as a cook in the navy on the continental establishment; that about the month of ------- A.D. 1779 he enlisted at L'orient, in France, on board the ship Bon Homme Richard, fitted out there for the Continental service by Doctor Franklin, under the command of Captain Paul Jones;

that he served on board the said ship nearly a year, and all together about four years, that in the engagement with the English frigate Serapis, the deponent lost his right leg, and after his recovery he was made cook; that after the engagement between the Bon Homme Richard and the British frigate Serapis, the deponent, with the rest of the crew, went onboard of the Serapis, their own ship having been sunk, and went into Holland, thence to France, where they were put on board the Arial, hired by Doctor Franklin, and returned to America, and shipped again as captain's cook on board of the Trumbull frigate, Captain Nicholson, where he served about a year, where he was taken prisoner, and remained so at New York, four or five months.

Author's note – The Trumbull was captured on the 8th of August 1781 by the British frigate Iris.

That he was exchanged, returned to Philadelphia, and shipped on board the Continental ship Duke Lusaune, Captain Green, and served in her as cook until the peace.

Author's note – Correct spelling of Duke Lusaune is Duc De Lauzun.

That by reason of his reduced circumstances in life, he is in need of assistance from his country for support, that he is a resident citizen of the United States.

Author's note – James Mc Kinzey's pension application was approved and he received $8 per month for 1 year as a seaman.

There is no mention of his death.

John Pickens

It's not unusual to come across a pension application that states that the applicant saw General Washington, but it is very rare that the applicant has a verbal encounter with him. And in this application, we have just that. An additional plus, is that this is a very well written and interesting pension claim.

The State of Ohio, Meigs County

On this fifth day of February A.D. One Thousand Eight Hundred & Thirty Four personally appeared before me John C. Bestow an Associate Judge of the Court of Common Pleas for said County John Pickens a resident of Meigs County in the State of Ohio aged Eighty Two years who being first duly sworn according to law, doth on his oath make the following declaration in order to obtain the benefit of the Act of Congress passed June 7th 1832. That he entered the service of the United States under the following named officers served as herein stated to wit, he entered the service as a volunteer at first for the term of one year early in the spring of the year (as near as he can recollect) 1776 at Carlisle in the State of

Pennsylvania in the company commanded by Capt. Adams in the Pennsylvania Line. Marched either on the 15th or 17th of March from Carlisle to Ticonderoga where the troops continued for a time & then returned to a place called Three Rivers.

There his officers were all killed by the Indians & the company to which the deponent belonged returned to Bergen near N. York three days after the Battle of Long Island & was with the greater part of his company placed under the command of Capt. McElhatton in the Battalion commanded by Col. Frederick Watt. Was with said company & Battalion which was called the "Flying Camp" with the Army on its retreat through N. Jersey under command of Genl. Irwin – At the time of the attack on Trenton we lay below Trenton & above Bordentown on the Pennsylvania side. After the Battle of Trenton crossed the Delaware was not in the engagement at Princeton – Although there the same day afterwards marched to Morristown where we continued during the winter & until his year had expired when he received a discharge signed by Col. Frederick Watt which he kept till about 20 years since when his house, situated on the Ohio, in this county, was destroyed by high water together with most of his property & his discharge. After his discharge he returned to Shearman's Valley his place of residence about 18 miles from Carlisle. Sometime in the summer following (1777) Deponents younger brother was drafted to serve a tour in the Militia of three months & deponent volunteered for the purpose of being with & taking care of his brother who had been unaccustomed to hardships or to being from home & at the expiration of the 1st three months volunteered for three months longer making 6 months in the company commanded by Capt. Fisher in Col. Dunlaps Battalion of Riflemen. While out on this tour of Militia the British landed at Elk River. His Regt. Was present & there was some skirmishing but the enemy landed under the cover of their shipping – was engaged in skirmishing with the enemy at Conch's Bridge.

When our Army lay at a place which he thinks was called Whitely Col. Dunlaps Battalion was stationed 5 miles toward the enemy he then acted as Sergeant commanded a Picket Guard two miles from the Battalion near a place called Iron Hill. While on duty Genl. Washington with some officers accompanied by a troop of Cavalry rode up. Deponent hailed them & ordered them to stand. Genl. Washington who was in a plain dress asked if deponent did not know him – to which he replied I know no man while on duty – Genl. Washington asked why they might not pass to which deponent replied the enemy are stealing a march, they are advancing – listen & you will hear them. Silence was ordered & they were distinctly heard. Genl. Washington asked deponent his name & what corps he belonged, noted (as he supposed) his answer & rode off at full speed. Soon after, the enemy appeared in sight – gave the alarm & joined his battalion, skirmished with the enemy that day, at night joined the main army. At roll call next morning, he was called out in front to the officers & was fearful he had not done exactly as he ought the day before, but to his great joy received the praise of Genl. Washington in presence of several officers & a present from him for his fidelity. He was also in the battles of Brandywine, Germantown, & at a place called White House. At the latter place the two armies soon separated on account of the excessive rain, After the expiration of his six months he received a discharge which he thinks was signed by Genl. Irwin & which was destroyed as before related.

He returned to Shearman's Valley. In the spring of the following year (1778) he enlisted in Carlisle Pennsylvania during the war in the company commanded by Capt. Buchanan, he was soon transferred to a company of Riflemen the names of the officers he does not recollect, the brigade commanded by Genl. Wayne & was in the Battle of Monmonth where he was wounded in the hip & removed with a Sergeant command of 21 men to Hackensack Ferry where said detachment took a sloop loaded with provisions. He stayed at this place about two months & being still unable to walk was taken to Trenton in one of the wagons that took flour from this sloop to Trenton. From thence was conveyed to Philadelphia & remaining still feeble sent to his friends who conveyed him on a bed in a wagon home to Shearman's Valley where he remained unable to walk except with a crutch for near 5 years & wholly unable to do military duty. Still painful at times, he has no documentary evidence & knows of no person whose testimony he can procure to testify as to his services.

Author's note – John Pickens's pension application was approved and he received $60 per year for 1 year and 6 months of service as a private.

There is no mention of his death.

Edward Quigley

In this interesting claim from the 1818 Pension act, the applicant demonstrates the determination to keep on fighting for America, even after being captured and wounded several times.

State of Pennsylvania, Centre County

Before me the subscriber one of the Justices of the Peace in and for the County of Centre aforesaid, also Register for the probate of wills and granting Letters of administration, and one of the Judges of the Registers Court in and for the County of Centre aforesaid personally appears Edward Quigley aged Sixty Six years, resident in Bald Eagle Township in said County who being first duly sworn according to law doth on his oath make the following declaration in order to obtain the provision made by the late act of Congress entitled an "act to provide for certain persons engaged in the land and naval services of the United States in the Revolutionary War." That he the said Edward Quigley enlisted in Lancaster County in the State of Pennsylvania on the 25th day of December AD 1775 in the Company Commanded by Capt. Joseph Habley of the 3rd Regiment of Infantry, that he continued to serve in the said Corps & in the service of the United States until the 12th day of November AD 1776 when he was taken prisoner by the British at Fort Washington, and continued prisoner until about the beginning of May following, when he was paroled and he again immediately joined the American Army in the Company commanded by Capt. Batein.

Author's note – Once paroled Edward Quigley was not supposed to rejoin the war on the side of America, per his parole agreement. If caught fighting for America by the British he could have been killed.

And continued in said Corps in the service of the United States for about fourteen months when he was drafted into the 4th Regiment in the company commanded by Capt. Graydon (as he thinks) that he continued in this company until after the Battle of Brandywine, that shortly after this he among others was sent under the command of Capt. McNeill to join the troops under General Knox in South Carolina, that he continued to serve in the service of the United States in the companies to which he was at different times drafted to for and during the full and term of the war, when he was discharged at Philadelphia from the company commanded by Capt. Simons of the 4th Reg. of Artillery commanded by Col. Porter, that he was in the Battles of Long Island where he was wounded by a bayonet in the right shoulder, and also wounded by a cut in the head within about a mile of Shades Ford on the Schuylkill, that at the Battle of Brandywine he was shot through the cheeks, cutting his tongue and knocking out some of his teeth, that he was also at the Battles of Germantown, Kings Mountain, Cowpens, at Sattons

Creek and in several other engagements too numerous to recite. But finally he was at the Capture of Lord Cornwallis and his army at Yorktown.

Author's note – Edward Quigley's pension application was approved and he received $8 per month.

There is no mention of his death.

Conclusion

I hope this sampling of Revolutionary War pension applications from the State of Pennsylvania has given you an insight into the men who fought for their independence.

Although there were plenty of interesting pension claims to choose from, I felt that this group of 40 covered the spectrum of service men performed during the war in Pennsylvania. I struggled with the thought of adding more, but decided these stories conveyed the sacrifices made during this war.

For a more detailed look into two men who fought for Pennsylvania, I encourage you to read my book about Captain Thomas Askey and Lieutenant Richard Gunsalus. The book chronicles their service through the entire Revolutionary War and expands on a lot of the military regulations and policy that governed them.

You can find information on all my books in the Author chapter of this book.

References

1. Pennsylvania Historical & Museum Commission
2. Pennsylvania State Archives series 1 volume V page 393
3. Pennsylvania State Archives series 1 volume VIII page 422/3
4. www.anamericanfamilyhistory.com
5. www.warnewjersey.com
6. www.wikipedia.org
7. www.wikitree.com
8. www.revolutionarywar.com
9. www.history.navy.mil

The National Archives State of Pennsylvania pension applications of;

Thomas Bull

Robert Carr

Gabriel Cory

Robert Covenhoven

David Criswell

Cornelius Dailey

Franz Dido

Peter Dooey

Hugh Drennan

Leonard Engler

Mathias Fisher

James Flack

William Gill

Michael Graham

Jacob Grist

James Guffey

James Guthrie

Barnet Hageman

Frederick Hain

John Hall

Joseph Harborn

James Hays

Jacob Hefflebower

George Heinish

John Hoge

James Huston

William Jenkinson

James Johnston

James Josiah

Frederick Kalehoft

John Kessler

James Knight

Thomas Laidley

Ephraim Lewis

William Lyons

Peter Mathews

Angus McCoy

James Mc Kinzey

John Pickens

Edward Quigley

About the Author

Ed Semler retired from the United States Coast Guard in December of 2007 with over 25 years of military service in both the United States Army and United States Coast Guard. In the United States Army he was an enlisted man and was honorably discharged as a Specialist Four (E-4). While in the United States Coast Guard he was enlisted, obtaining the rank of Master Chief Petty Officer (E-9), was commissioned as an officer, and retired as a Lieutenant (O-3E).

Fully retired he resides in Schulenburg, Texas with his wife Jana, a retired Air Force senior master sergeant. Please feel free to check out Ed's other books at www.edsemler.com or email him at mkcm378@gmail.com

He also has a YouTube channel www.youtube.com/@MKCMLT which is full of military related videos.

His other publications are;

"Around The World," a memoir of his 25 years of service as an officer and enlisted man in the U.S. Army and U.S. Coast Guard

"U.S. Coast Guard Cutter Sherman (WHEC-720) Circumnavigation Deployment 2001" which details the *Sherman's* historic circumnavigation of the globe and deployment to the Persian Gulf in 2001

"The Three Gunsallus Brothers" a story about fighting for Pennsylvania during the Civil War

"Sam Houston & Napoleon Bonaparte Meet On The Civil War Battlefield" a true story of the Walker brothers

"Thoughts On Being A Chief Petty Officer" a take on military leadership

"Fighting For Pennsylvania In The Early Years 1763 to 1783 – The Story of Captain Thomas Askey And Lieutenant Richard Gunsalus Of Cumberland County"

"Joe Semler Playing Baseball in the 1920's &30's"

"Alice Springs Australia Adventures In The 80's"

"Count On Us Coast Guard Cutter Dependable – Law Enforcement And Search & Rescue"

"United States Coast Guard Tragedies"

www.ingramcontent.com/pod-product-compliance
Lightning Source LLC
LaVergne TN
LVHW051728080426
835511LV00018B/2937